A TREATISE ON STATISTICAL PARADOXES

SLIUATION THAT SEEMS TO CONTRAVENE INTUITION

A TREATISE ON STATISTICAL PARADOXES

SLIUATION THAT SEEMS TO CONTRAVENE INTUITION

Ratna Raj Laxmi
Chetan Verma

KALPAZ

ISBN: 978-93-5128-132-0

Price: ₹ 750

First Published, 2015

Published by

Kalpaz Publications
C-30, Satyawati Nagar,
Delhi-110052
Ph. : 9212142040
E-mail: kalpaz@hotmail.com

Cataloging in Publication Data—DK
Courtesy: D.K. Agencies (P) Ltd. <docinfo@dkagencies.com>

Verma, Chetan, author.
A treatise on statistical paradoxes / Chetan Verma.
pages cm
Includes bibliographical references.
ISBN 9789351281320

1. Statistics. 2. Paradoxes. I. Title.

DDC 519.5 23

Contents

Preface

In this book some of the paradoxes from the history of statistics have been presented which have inspired the clarification of basic concepts and the introduction of major results. The term paradox is used in broad sense to mean an inconsistency, a counter to widely held notions, a misconception, a true statement that seems to be false, or a false statement that seems to be true. It is in these various senses that paradoxes have played an important role, in expressed in words by Bell.

"The mistakes and unresolved difficulties of the post in mathematics have always been the opportunities of its future."

With this broad sense paradoxes are considered fewer than two categories Statistics and Statistical Decision Theory. Efforts have been made to express these paradoxes in simple and understand able form.

Chapter 1

Introduction

HOW WONDERFUL THAT WE HAVE MET WITH PARADOX, NOW WE HAVE SOME HOPE OF MAKING PROGRESS: NIELS BOHR

Paradox in philosophy refers to a problem that has two or more conflict answer each generated by using the logic "correctly". In other words, a paradox generated obviously unaccepted conclusions which are derived by apparently acceptable reasoning process from apparently acceptable premises. Paradoxes arouse crises of through and provoke revolutionary advances. Paradox has been described as a truth standing on its head to attract attention. Undoubtedly, paradoxes captivate. They also cajole, provoke, amuse, exasperate and seduce. More importantly, they arouse curiosity, stimulate and motivate. A taste for paradox is central to the philosophies of Laozi, Heraclitus, Meister Eckhort, Hegel, Kierkegaard and Chesterton, among many others. Soren Kierkegaard (1985), for example, writes in the Philosophical Fragments, that

> "But one must think I'll of the paradox, for the paradox is the passion of through and thinker without the paradox is like the thinker without the paradox is like the lover without passion, a mediocre fellow. But the ultimate potentiation of every passion is always to will its own downfall and so it is also the ultimate passion of the understanding to will the

> collision, although in one way or another collision must become its downfall. This, then, is the ultimate paradox of thought, to want discover something that though itself cannot think."

Most logical paradoxes are known to be invalid arguments but are still valuable in promoting critical thinking. Some paradoxes have revealed errors in definitions assumed to be rigorous and have caused axioms of mathematics and logic to be re-examined. One example is Russell's paradox, which questions whether a "list of all lists that do not contain themselves" which shows that the notion of the set of all those sets that do not contain themselves leads to a contradiction, was instrumental in the development of modern logic and set theory.

In general, paradoxes can also serve a useful role in the classroom. The temporary confusion and insecurity which they may engender in students can be put to good use. Conflict and predicament are useful pedagogical devices. They may foster a positive attitude to getting stuck, provide and promote the opportunity to participate in debate and controversy over statistical or mathematical issues and promote the realization that statistics or mathematics often develops in this way. Teachers may gain a better appreciation of student's difficulties in coming to grips with concepts and results with which some of the greatest statisticians of all time struggled. Such concepts and results, while paradoxical and challenging at the time, become common place is subsequent generations. In the words of Kasner and Newman (1967).

> "The testament of science is so continually in a flux that the heresy of yesterday is the gospel of today and the fundamentalism of tomorrow."

Many paradoxes, in statistical theory, are involving functions. The concept of function originated in the early 18th century. Newton and Leibniz invented the calculus in the latter part of the 17th century. A function was viewed at different times as a formula, a curve, or arbitrary correspondence paradoxes turned up to dethrone one or another of these views of functionality. Even the very meanings of a formula as well as its scope changed over time and were often subjects of considerable controversy. For example, in an 1829 paper on Fourier series Dirichlet introduced the so called Dirichlet fuctions.

$$D(x) = \begin{cases} 1 & \text{if } x \text{ is rational} \\ 0 & \text{If } x \text{ irrational} \end{cases}$$

This function was neither a formula nor a curve. It was a new type of function, described by a correspondence. It was the first of many functions which came to called "pathological", but not for very long. At the end of the 19th century Baire extended the notion of formula. To him it meant an expression obtained from a variable and constants by a iteration of additions, multiplications and the taking of limits. He called such function analytically representable and showed that Dirichlet function is of type

$$D(x) = \lim_{m \to \alpha} \lim_{n \to \alpha} \cos(m! \pi x)^n$$

Thus the "pathological" Dirichlet function becomes a "tame" analytically representable function. The main question is arises is analytic represent ability of function? That is, are functions which are not analytically representable? Yes and no. If you are a formalist, you can show by a counting argument that the set of analytically representable function has cardinality 'c' while the set of all functions has cardinality 2^C. Thus there are uncountable many functions which are not analytically representable. But no one has given a constructive example of even one. Although the concept of continuity is nowadays fundamental in Statistics or Mathematics, its modern definition was not formulated until the 19th century, about 150 years after the invention of the calculus by Newton and Leinig. In the 18th century, Euler did define a notion of continuous function was one given by a single expression, while a function given by several expression was considered discontinuous. For example, to Euler the function

$$f(x)=\{\begin{matrix} x & x>0 \\ -x & x\leq 0 \end{matrix}$$

was discontinuous, while the function comprising the two branches of a hyperbola was considered continuous, since it is given by the single expression

$$f(x)=\frac{1}{x}$$

In an 1821 work Cauchy initiated a reappraisal and reorganization of the foundation of 18th century calculus. In this work Cauchy defined continuity essentially as we understand the concept today, although he used the then-prevailing language of infinitesimals rather than the now accepted $\in$ - δ formulation given by Weierstrass in the 1950's. In the former case continuity was a global property while in the latter case it was a local property. But the concept of continuity proved to be very subtle and was not completely understood even by Cauchy and his contemporaries of the early to mid-19th century. For example, Chaushy proved that an infinite sum of continuous function is a continuous function. This, of course, is incorrect. A counter example was given by Abel in 1820, it is essentially the series

$$\sum_{n=0}^{\infty}\frac{Sin(2n+1)^{x}}{2n+1}$$

Which is discontinues for all x = kπ, k = 0, ± 1, ± 2,........

The error in Cauchy does prove resulted from his failure to distinguish between convergence and uniform convergence of a series of functions. In a modern development, Schwartz and Sobolev showed in the 1940's that even continuous function is, indeed, differentiable. For example,

$$f(x)=\begin{cases} 1 & x>0 \\ \frac{1}{2} & x=0 \\ 0 & x<0 \end{cases}$$

then

$$f'(x) = \begin{cases} 0 & x \neq 0 \\ \infty & x = 0 \end{cases}$$

which is the Dirac delta function $\delta(x)$. As this example shows, there are even discontinuous function which are differentiable, a shocking realization for statistician or mathematicians of the second half of the 19th century.

Furthermore, in most of the statistical paradoxical situation set theory is involving, for example, Russell's paradox. During nearly the last three decades of the 19th century Cantor developed many important set theoretic ides using an intuitive notion of set. Eventually his concept proved inadequate and led to paradoxes. Perhaps the best known is Russell's paradox of 1902. Let

$$R = \{x : x \notin x\}$$

then $R \in R$ iff $R \in R$ This paradox had a profound effect on a number of statisticians. It devastated the logician Frege, who had just completed a two-volume treatise on the foundation of arithmetic which relied on set-theoretic notions and learning of Russell's paradox, he lamented.

> "A scientist can hardly meet with anything more undesirable than to have the foundation given way just as word is finished. I was put in this position by a letter from Mr. Bertrand Russell when the work was nearly through the press."

On the other hand, the paradoxes of set theory had positive effects. In particular, they provoked statisticians to precise meaning to notion of set by devising various axiomatizations of set theory, e.g. the Zermelo-Fraenkel axioms, the Russell and Whitehead theory of types, the Gochel-Bermays system. Although such axiom-system avoided the known paradoxes, they did not guarantee that new ones would not emerge. As Poincare put it picturesquely

> "We have put a fence around the heard to protect it from the wolver but we do not know whether some wolver were not already within the fence."

The two paradoxical results from Cantor's work in set

theory are, infinite comes in different sizes, in fact, in infinitely many different sizes. The second result comes from juxtaposing the following two quotations

(a) "Later generations will regard Mengenlehre (set theory) as a disease from which one has recovered."

(b) "No one shall expel us from the paradise which Cantor created for us."

In general, there is another class of paradox in Statistics, most of the paradoxes relate to decision-making. The decision theory in statistics is mainly concerned with identifying the values, uncertainties and other issues relevant in a given decision, its rationality and the resulting optimal decision. The paradox relates to decision-making and it was identified by Triantaphyllou and Mann (1989), it is generally known as "Decision-making paradox." The realization for decision-making paradoxes come from the rather straight forward observation that there are numerous decision-making methods each one of which claims be the best one. Furthermore, often these methods may yield different results when they are fed with exactly the same decision problem and data. In other words, decision theory concerned with identifying the best decision to take assuming an ideal decision maker who is fully informed, able to compute with perfect accuracy and fully rational. In practice, there are situations in which best is not necessarily the maximal, optimum may also include values in addition to maximum, but within a specific or approximate range. The practical application of this prescriptive approach i.e. how people ought to make decisions, is called decision analysis and aimed at finding tools, methodologies and software to help people make better decisions. The choice of the decision under uncertainty represents the heart of decision theory. The procedure now referred to as expected value and the idea of this is that, when faced with a number of actions, each of which could give rise to more than one possible outcome with different probabilities, the rational procedure is to identify all possible outcomes, determine their values and the probabilities that will result from each course of action and multiply the two to given an expected value. The action to be chosen should be the one that gives rise to the highest total expected value. But, in the 20th century, interest was reignited by Abraham Wald's in 1939 paper pointing out that the two central procedure of sample-distribution based statistical-theory, namely hypothesis testing and parameter estimation are special cases of general decision problem. Wald's paper renewed and synthesized many concepts of statistics theory, including loss function, risk functions, admissible decision rule, antecedent distribution, Bayesian procedures and minimax procedures.

Statistical decision theory is concerned with the making of decisions in the presence of statistical knowledge which sheds light on some of the uncertainties involved in the decision problem. In Bayesian procedures, most of the time we assume that these uncertainties can be considered to be unknown numerical quantities and will represent by θ. As an example, consider the situation of a drug company deciding whether or not to market a new relieve. Two of the many factors affecting its decision are the proportion of people for which the drug will prove effective (θ_1) and the proportion of people for which the drug will capture(θ_2). Both θ_1 and θ_2 will be generally unknown, though typically experiments can be conducted to obtain statistical information about them. This problem is one of decision theory in that the ultimate purpose is to decide whether or not to market the drug, how much to market, what price to change, etc. Classical statistics is directed towards the use of sample information in making inferences about θ. These classical inferences are made without regard to the use to which they are to be put. But sometimes conclusions are drawn according to their prior beliefs or prior probabilities. The approach to statistics which formally seeks to utilize prior information is called Bayesian analysis. Because of their wide range of applicability of Bayesian approach most of the paradoxes can be resolved using this approach.

Paradoxes in the subject of Statistics and Statistical Decision Theory altogether with broad sense are considered here. Efforts have been made to express these paradoxes in the simple and understandable form. The book intends to serve a number of collogues and students for the asterisks marked as.

It is in these various senses that paradoxes have played an important role, in expressed in the words

> "One of the endlessly alluring aspects of mathematics is that its thorniest paradoxes have a way of blooming into beautiful theories."
>
> ---Davis

Chapter 2

Abelson Paradox

EFFECT SIZE MAY NOT BE INDICATIVE OF PRACTICAL MEANING.

Abelson's paradox is an applied statistical paradox identified by Robert P. Abelson (1985). According to him, concerning a single major league at bat, the percentage of variance in batting performance attributable to skill differentials among major league baseball players, can be calculated statistically. The statistically appropriate calculation is seriously discrepant with intuitions about the influence of skill in batting performance. He discussed this paradoxical discrepancy in terms of habits of thought about concept of variance explanation.

In his study Abelson found that "the percentage of variance in any single batting performance explained by batting skill is about one third of 1%". Due to this result his personal intuition was jarred, which seems much too small. To check it, he prepared unambiguous questionnaire and circulated to all graduate students and faculty in the department of Psychology at Yale University. This group was chosen not simply for convenience, but because they would be familiar with the concept of variance explanation. Respondents were asked to refrain from answering if they knew nothing about baseball or the concept of variance explanation. Participants' were asked to imagine a time at bat by an arbitrarily chosen major league baseball player, and to estimate the percentage of the variance in whether or not the batter gets a hit is attributable to skill differentials between batters.

He studied the median of the 61 estimates of the variance attributable to skill was 25% and an overestimate of the calculated estimate by factor of 75. The estimate of over 90% of the sample was too high by a factor of at least 15. Only one person gave an underestimate. Abelson also posed the skill variance question to colleagues outside of Yale University and commonly received answers around 20% or 30%. This suggests parameters such as μ_B= 0.270 and μ_B= 0.025 The outcome of the statistical calculation, 0.3%, is indeed surprising.

Another polemic on the paradox is to look for flaws in the statistical calculation. Abelson gave an equation, the omega-squared ratio for proportion of variance attributable to skill, is given by

$$\omega^2 = \frac{\sigma_B^2}{\sigma_B^2 + \sigma_B^2} = \frac{\sigma_B^2}{\sigma_B^2 + (1-\mu_B) - \sigma_B^2} = \frac{\sigma_B^2}{m_B(1-\mu_B)}$$

where μ_B is the mean of the distribution and σ_B be the standard deviation and σ_e^2 is the within batter variance.

He considered the sensitivity of the above equation to variation in the parameters σ_B and μ_B. The term $\mu_B(1—\mu_B)$ does not change appreciably with small variation in μ_B. He saw that if μ_B = 0.265 or 0.260 or 0.275 rather than 0.270. The ratio is more sensitive through, to variations in σ_B. If lifetime batting averages are taken as more indicative of true ability than season-by-season averages their standard deviation would be used for σ_B. He calculated mean life average from the data in James's (1983) baseball abstract which was 0.268 and standard deviation of lifetime average was 0.021. In this practical situation he raised a question, could the equation even give a large value for ω^2. He found out an answer, yes, if every batter batted either 1.000 or 0.000 (i.e. either perfect or perfectly awful), then one would expect then $\sigma_B^2 = \mu_B (1- \mu_B)$ and $\omega^2 = 1$. This extreme situation contrasts sharply with reality. Indeed, a way to understand the paradox is to realize that in the major leagues, skills are much greater than in the general population. However, even best batters make outs most of the time. This is the paradox he talked about in his study.

He tried to find out the answer of some questions such as what does the baseball paradox suggest for the usual standards for conceptualizing variance explanation? If one-third percent indicates such a trivial degree of explanation variable in baseball? Or should one instead be more suspicious of variance explanation as an index of systematic influence and revise the notions surrounding less than 1% of variance explanation? The answer lies in the type of example that he gave in his study. First, the individual batter's success is appropriately measured over a long season, not by the individual at bat. Second, a team scores runs by conjunctions of hits, so a team with many high-average batters is more likely to stage rallies than a team with many low average batters. Thus, team success over a long season is influenced by average batting skill far more than is individual success in the single at bat because the effects of skill cumulate, both within individuals and for the team as a whole. The statistical effects of cumulating are well known, although they are usually discussed in methodological contexts. In sum, the large intuitive overestimation of the variance in batting outcome explained by skill is not simply an error in the appreciation of statistics. It reflects an intuition that skill does matter. Indeed it does, in the long run, albeit not very consequentially in the single episode. The baseball paradox is thus a model for similar paradoxes that may arise in psychological contexts.

Bibliography

(1) Abelson, Robert P. (1985). A variance Explanation Paradox: When a Little is a Lot, American Psychological Association,

Vol. 97, pp 129-133.

(2) Cohen, L.J. (1981). Can human irrationality be experimentally demonstrated? The Behavioral and Brain Science, Vol. 4, pp 317-331.

(3) James, B. (1983). *The Bill James baseball abstract,* New York, Ballantine Book.

(4) Langer, E.J. (1975). The illusion of control, Journal of Personality and Social Psychology. 32, pp 311-328.

Chapter 3

Abilene Paradox

PEOPLE CAN MAKE DECISIONS BASED NOT ON WHAT THEY ACTUALLY WANT TO DO, BUT ON WHAT THEY THINK THAT OTHER PEOPLE WANT TO DO, WITH THE RESULT THAT EVERYBODY DECIDES TO DO SOMETHING THAT NOBODY REALLY WANT TO DO, BUT ONLY WHAT THEY THOUGHT THAT EVERYBODY ELSE WANTED TO DO.

In an Abilene paradox, a group of people collectively decide on a course of action that is counter to the preference of many of the individuals in the group. It involves a common breakdown of group and therefore does not raise objections. A common phrase relating to the Abilene paradox is a desire to not "rock the boat". The Abilene paradox was introduced by management expert Harvey (1974). The name of the phenomenon comes from an anecdote in the article which Harvey uses to elucidate the paradox.

> "The July afternoon in Coleman, Texas was particularly hot. In addition, the wind was blowing fine. But the afternoon was still tolerable, even potentially enjoyable. There was a fan going on the back parches, there was cold lemonade and there was entertainment, Dominoes. The family was comfortably playing dominoes on a parch making an agreeable Sunday until the father-in-law suggested that "let's get in the car and go to Abilene and have dinner at the Cafeteria." The wife said, "Sounds like a great idea." The husband, despite having reservations because of long drive

and dust storm and thought that his preferences must be out-of-step with the group and replied "Sounds good to me" and added "I just hope your mother wants to go."
The Mother-in-law suddenly said "Of course I want to go. I haven't been to Abilene in a long time." The drive was hot, dusty and long. When they arrived at the cafeteria, the food was as bad as the drive. They returned home four hours later, exhausted. After a long time of silence the husband said "It was a great trip, wasn't it?" The mother-in-law said with some irritation "well, to tell the truth, I just went along because the three of you were so enthusiastic about going. I wouldn't have gone if you all hadn't pressured me into it." The husband said, "What do you mean 'You all'?" "Don't put me in the... 'You all' group...culprits." The wife looked shocked "I just went along to be sociable and to keep you happy. I would have had to be crazy to want to out in the heat like that." The father-in-law then said "I just thought you might be bored. You visit so seldom I wanted to be sure you enjoyed it. I would have preferred to play nether game of dominoes and eat the leftovers in the icebox."

This anecdote was also made into a short film for management education. The Abilene paradox is similar to groupthink but differs in significant ways, including that in group think has reached. In the Abilene paradox, the individuals acting contrary to their own wishes are more likely to have negative feelings about the outcome.

The corollary to the Abilene paradox is that the inability to cope with agreement, not conflict, is central to organization dysfunction. According to Harvey (1974) experiences, the Abilene paradox can occur the bedroom, boardroom, oval room and the airport. He then cites several instances, illustrating example.

The Bedroom: A young woman approached Harvey, telling him how depressed she was because she was supposed to get married at the end of the week, but she didn't love her fiancé. In fact, she didn't even like him. She felt compelled to marry him because she had already accepted the engagement ring, he was a good guy, she didn't want to break his heart, all the guests were on their way and she didn't want to back out and cause her mother to have a heart attack. Later on, her fiancé approached Harvey and said he was depressed because he was to be married at the end of the week and he didn't love or even like her! But he was going through with it because he had already proposed to her, she was a good woman, he didn't want to break her heart, all the guests were on their way and he didn't want to curse his fiancé's mother to have a heart attack. The outcome was that they got married, although neither wanted to and eventually they divorced.

The Boardroom: A company was planning to undertake a project that would, if it failed, bankrupt the company. Although each key employee believed the project was doomed to failure, no one said anything because each thought that everyone else knew better or knew more than he did. Eventually after the company started their journey to Abilene, Harvey stepped in and voiced his opinion which was merely a representation of every else's opinion.

The Oval Room: The Watergate bugging turned out to be a trip to Abilene when questioned as to why the bugging occurred, each participant testified that he didn't think it was a good idea and couldn't see any valid reason to do it. It was illegal immoral and served no political purpose. Yet they did it. Each participant want along with the plan thinking that the others all wanted to and that they knew something the others didn't. This was an example of a trip to Abilene with the whole country going along for the ride!

The Airport: A man on a business trip begins to fantasize that, upon his return; his wife will be waiting to great him while wearing her negligee and holding a bottle of champagne. She will tell him her mother is watching the kids and that they have a reservation at a nearby hotel so they can have a romantic weekend together. Her fantasy, while waiting for him, is that he will arrive with a dozen roses and announce to her that he has made reservation at a fancy restaurant and an expensive hotel so they can spend a romantic weekend together. But, upon actually greeting each other, they get into an argument and end up spending the weekend apart.

In general, the phenomenon is explained by social psychology theories of social conformity and social influence which suggest that human being are often very averse to acting contrary to the trend of a group [Levi (2010), Vasu et al. (1988)]. But it include the some concept of decision making. According to Harvey, the phenomenon may occur where individuals experience action- anxiety, stress concerning potential negative attitudes to them of the group if they do not go along. This action-anxiety arises from what Harvey termed "negative fantasies" unpleasant visualizations of what the group might say or do if individuals are honest about their opinions, where there is real risk of displeasure and negative consequences for not going along and the individual has separation anxiety, fearing exclusion from the group.

Harvey (1974) can stated the Abilene paradox succinctly as, organizations frequently take actions in contradiction to the data they have for dealing with problems and as a result, compound their problems rather than solve them. Like all paradoxes, this paradox, deals with absurdity. On the surface, it makes little sense for organization, whether they are couples or companies, bureaucracies or governments, to take actions that are diametrically opposed to the data they possess for solving crucial organizational problem. Such actions are particularly absurd since they tend to compound the very problems they are designed to solve and thereby defeat the purposes the organization is trying to achieve.

The paradoxes are generally based on a logic or rationale different from what we understand or expect. Discovering the different logic not only destroys the paradoxical quantity but also offers alternative ways for coping with similar situations. Therefore, part of the dilemma facing an Abilene-bound organization may be the lack of a map, a theory or model that provides rationality to the paradox. Harvey (1974) provide such a map which will be developed by examining the underlying psychological themes of the profit-making organization and bureaucracy and it will include, action anxiety, negative fantasies and Psychological reversal of risk and certainty.

Bibliography

(1) Harvey, J.B. (1974). The Abilene paradox: The management of agreement, Organization Dynamics, 3, pp 63-80.

(2) Levi, Daniel (2010). *Group Dynamics for teams,* SAGE Publication, pp 108-109, ISBN 978-1-4129-7762-3.

(3) McAvoy, J.B.T. (2006). Resisting the change to user stories: A trip to Abilene, International Journal of Information Systems and change Management, 1(1), pp 48-61.

(4) McAvoy, J.B.T. (2007). The impact of the Abilene Paradox on double-loop learning in an agile term, International and Software Technology, 46(6), pp 552-563.

Chapter 4

Accuracy Paradox

PREDICTIVE MODELS WITH A GIVEN LEVEL OF ACCURACY MAY HAVE GREATER PREDICTIVE POWER THAN MODELS WITH HIGHER ACCURACY.

The accuracy paradox for predictive analytics states that predictive models with a given level of accuracy states that predictive models with a given level of accuracy may have greater predictive power than models with higher accuracy when accuracy is defined as the percentage of correct predictions. It may be better to avoid the accuracy metric in favor of other metrics such as precision and recall. Before exploding much about accuracy paradox we have to know about predictive analytics, it pledge a variety of techniques from statistics, modeling, machine learning and data mining. That analyzes current and historical facts to make predictions about future. The term predictive is used to mean predictive modeling, "scoring" data with predictive model and forecasting. Predictive models analyze past performance assess how likely a customer is to exhibit a specific behavior in order to improve marketing effectiveness. This category also pledges models that seek out subtle data patterns to answer question about customer performance, such as fraud detection model.

Accuracy is often the starting point for analyzing the quality of predictive model, as well as a candid criterion for prediction. Accuracy measures the ratio of correct predictions to the total number of cases evaluated. It may seem candid that ratio of correct predictions to cases should be a key metric. A predictive model may have high accuracy but may be useless. In an example predictive model for an insurance fraud application, all cases that are predicted as high-risk by the model will be investigated. To evaluate the performance of the model, the insurance company has created a sample data set of 5000 claims. All 5000 cases in the validation sample have been carefully checked and it is known which cases are fraudulent. To analyze the quality of the model, the insurance uses the table of confusion, it is a table with two rows and two columns that reports the number of false positive; false negatives, true positives and true negatives. The definition of accuracy, the table of confusion for model MF_1 and calculation of accuracy for MF_1 is given by

$$Accuracy = \frac{N_{TN} + N_{TP}}{N_{TN} + N_{FP} + N_{FN} + N_{TP}}$$

where N_{TN} is the number of true negative cases

N_{FP} is the number of false positive cases

N_{FN} is the number of false negative cases

N_{TP} is the number of true positive cases

The table of confusion for fraud model MF_1 is given by

	Predicted Negative	Predictive Positive
Negative cases	4850	75
Positive cases	25	50

$$Accuracy = \frac{4850 + 50}{4850 + 75 + 50 + 100} = 0.98$$

$$Accuracy\ Percentage = 98\%$$

With an accuracy of 98% model MF_1 appears to perform fairly well. The paradox lays the fact that accuracy can be easily improved to 98.5% by always predicting "no fraud".

The table of confusion and the accuracy for this trivial "always predict negative" model MF_2 and accuracy of this model are shown below

	Predicted Negative	Predictive Positive
Negative cases	4925	0
Positive cases	75	0

$$Accuracy = \frac{4925+0}{4925+75+0+0} = 0.985$$

$$Accuracy\ Persentage = 98.5\%$$

Model MF_2 reduces that rate of inaccurate predictions from 2% to 1.5%. This is an apparent improvement of 25%. The new model MF_2 shows fewer incorrect predictions and markedly improved accuracy, as compared to the original model MF_1, but is obviously useless. The alternative model MF_2 with higher accuracy does not offer any value to the company for preventing fraud. The less accurate model is more useful than the more accurate model. Model improvements should not be measured in terms of accuracy gains. It may be going too far to say that accuracy is irrelevant; but caution is advised when using accuracy in the evaluation of predictive models.

Bibliography

(1) Abma, B.J.M (2009). Evaluation of requirements management tool with support for traceability-based change impact analysis, unpublished master thesis summited to University of Twente, Netherland, pp 86-87.

(2) Bruckhaus, Tilmann (2007), *The Business Impact of Predictive Analytics*, IGI Global, pp. 118-119, ISBN 978-1-59904-252-7.

(3) Zhu, Xingquan (2007), *Knowledge Discovery and Data Mining: Challenges and Realities*, IGI Global, pp. 118-119, ISBN 978-1-59904-252-7

(4) http://www.bestthinking.com/thinkers/dennis-lendrem?tab=blog&blogpostid=18769

Chapter 5

Apportionment Paradox

SOME SYSTEMS OF APPORTIONING REPRESENTATION CAN HAVE UNINTUITIVE RESULTS DUE TO ROUNDING

An apportionment paradox exists when the rules for apportionment in a politiçal system produce results which are unexpected or seem to violate common sense. To apportion is to divide into parts according to some rule, the rule typically being one of proportion. Certain quantities like milk can be divided in any proportion, whereas other, such as animals cannot. Only the whole numbers will do. In the latter case, there is an inherent tension between our desire to obey the rule of proportion as closely as possible and the constraint restricting the size of each portion to discrete values. This result, at time, in unintuitive observations or paradox. Several paradoxes related to apportionment, also called fair division, have been identified.

The Alabama paradox was the first of the apportionment paradox, discovered in 1880. Apportionment paradox first surfaced after 1870 census in US with 270 members in the house of representatives, Rhode Island got 2 representatives but when the house size was increased to 280, Rhode Island lost a seat. After the 1880 census C.W. Seaton (chief clerk of U.S. census office) computed apportionment for all house size between 275 and 350 members. He then wrote a letter to congress pointing out that if the house of representatives had 299 seats, Alabama would get 8 seats but if the House of Representative had 300 seats,Alabama would get 7 seats. There was more to come, when Oklahoma become a state in 1907, a recompilation of apportionment showed that the number of seats due to other states would be affected even though Oklahoma would be given a fair share of seats and the total number of seats increased by that number. The method for apportionment used during this period, originally put forth by Alexander Hamilton but not adopted until 1852, was as follows:

> "Suppose that there are m states with populations $P_1, P_2, \ldots\ldots\ldots, P_m$ and n seats to distribute. Let $P = \sum_{i=1}^{m} P_i$ be the total population and let $p_i = \frac{P_i}{p}$ be the relative population of state i, i.e., its proportion of the total population. Calculate $\mu_i =$ P_i p, this is the real number that would give exact proportionality. First round these down and give (μ_i) seats to state i. The sum of these numbers is almost always less than n, the exception is when all μ_i happen to be integers and the remaining seats are given to the states with largest remainders μ_i. In other words, μ_i is rounded up for the states with largest remainders and the number of states that are rounded up are determined so that the total number of seats becomes n.

The method is simple and intuitive and it does not bias against small or large states. However, in 1881 it was discovered that this method has a surprising and unwelcome behavior when the total number of seats is changed. It can happen that some state gets less representatives when the total number is increased. More precisely, using the population figures from the 1880 census, a total of 299 seats would give 8 to Alabama, but a total number of 300 seats would give only 7 to Alabama State. This counter intuitive behavior got the name Alabama Paradox.

The Alabama paradox is mathematically not strange, once it has been noticed. Consider three states A, B and C. If we increase n to n + 1, the number $\mu_i = p_i n$ is increased by p_i. If, for example, state C is small and states A and B large, then μ_C increases less that μ_A and μ_B. For simplicity suppose that none of these numbers passes an integer, so the integer parts (μ_i) remains the same for n and n + 1 for all three states, then their remainders $e_i = \mu_i - (\mu_i)$ increase by p_i and it may happen that the remainders ρ_A and ρ_B both are smaller than ρ_C when we distribute n seats, but that both become larger than ρ_C when we increase η to $\eta + 1$. If furthermore C had the smallest remainder that was rounded up, then the result is that C loses one seat while A and B gain one each. Of course, the description so for is purely deterministic but the question arises that "what is the probability that it occurs if n is increased to n + 1?"

State	Population	μ_i	Seats
A	53	5.30	5
B	33	3.30	3
C	14	1.40	2
Sum	100	10.00	10

State	Population	μ_i	Seats
A	53	5.83	6
B	33	3.63	4
C	14	1.54	1
Sum	100	11.00	11

The Population Paradox: An increase in a state's population can cause it to lose a seat. The population paradox was discovered around 1900, when it was shown that a state could lose seats in the House of Representatives as a result of an increase in its population. Hamilton's method is used to resolve this paradox.

The New State Paradox: Adding a new state with its fair share of seats can affect the number of seats due other states. The New State Paradox was discovered in 1907 when Oklahoma became a state. Before Oklahoma became a state, the House of Representatives had 386 seats. Comparing Oklahoma's population to other states, it was clear that Oklahoma should have 5 seats so the House size was increased by five to 391 seats. The intent was to leave the number of seats unchanged for the other states. However, when the apportionment was mathematically recalculated, Maine gained a seat (4 instead of 3) and New York lost a seat (from 38 to 37). The method used at the time was again Hamilton's method.

The Quota Paradox: The percentage of state's population times the house size is called the quota for that state, the lower quota is the quota rounded down and the upper quota is the quota rounded up. It can happen that a state receives a number of seats which is smaller than its lower quota.

By choosing a random number n we mean choosing n uniformly at random from $\{1,2,3,\ldots\ldots,n\}$ for some large integer N and then taking the limit as $n \to \alpha$. Thus, let S_i (n) be the number of seats n, by one seat at a time, from 1 to n and let v_i (N) be the number of times that state i suffers from the Alabama paradox i.e., the number of $n < N$ such that

$$s_i(n+1) < s_i(n)$$

If $\frac{v_i(N)}{N}$ converges to some value q_i as $N \to \alpha$, then the limit q_i is the probability that state i suffers from the Alabama paradox [Janson and Linusson (2012)]. In order to calculate this limit and show that it exists, we will make one mathematical simplification. Consider a set $\{x_1, x_2, \ldots\ldots, x_K\}$ of real number is linearly independent over Q if there is no relation i.e.

$$a_1x_1 + a_2 x_2 + \dots\dots\dots + a_K x_K = 0$$

with all coefficient a_i rational and not all 0.

Several paradoxes related to apportionment paradox have been identified. The population Paradox, New States Paradox and Quota Paradox the examples which are also in the list.

Bibliography

(1) Balinski, M. and Young H. Peyton (1982). *Fair Representation: Meeting the Ideal of One man, One vote,* Yale University Press. ISBN 0-300-02724-9.

(2) Janson, Svante and Linusson, Svante (2012). The probability of the Alabama paradox, Journal of Applied Probability, 49, pp 773-794.

Chapter 6

Arrow's Paradox

GIVEN MORE THAN TWO CHOICES, NO SYSTEM CAN HAVE ALL THE ATTRIBUTES OF AN IDEAL VOTING SYSTEM AT ONCE.

In decision theory, in social choice theory, Arrow's impossibility theorem (Arrow Paradox) , the General Possibility theorem, states that, when voters have three or more distinct alternatives or options, no rank order voting system can convert the ranked preferences of individuals into a community wide ranking while meeting with a specific set of criteria. These criteria are known as unrestricted domain, non-dictatorship, Pareto efficiency and independence of irrelevant alternative respectively. The paradox is named after economist Kenneth Arrow, who demonstrated the paradox in his Ph.D. thesis and popularized it in his 1951 book "**Social Choice and Individual Values.**"

The need to aggregate preference occurs in many disciplines, where one attempts to find an economic outcome which would be acceptable and stable. In decision theory, where a person has to make a rational choice based on several criteria, and almost naturally in voting systems, which are mechanisms for extracting a decision from a multitude of voter's preferences. The framework for Arrow's paradox assumes that we need to extract a preference order on a given set of options. Each individual in the society gives a particular order of preferences on the set of outcomes. The theorem or paradox considers the properties like non-dictatorship and more important independent of irrelevant alternatives. For understanding these properties, Let A be the set of outcomes, *N* be a number of voters or decision criteria, and *L(A)* be the set of all full linear orderings. Then social welfare function or preference aggregation rule is a function

$$F : L\,(A)^N \;\rightarrow\; L\,(A)$$

which aggregate voter's preferences into a single preferences order on A. The N-tuple $R_1, R_2, \ldots\ldots, R_N$ of voter's preferences is known as preference profit. In its strongest and simplest form, Arrow's paradox states that whenever the set A of possible alternative has more than 2 elements, then the following three conditions become incompatible.

Pareto efficiency: If alternative 'a' is ranked above 'b' for all orderings $R_1, R_2, \ldots\ldots, R_N$, then 'a' is ranked higher than 'b' by $F\,(R_1, R_2, \ldots\ldots, R_N)$.

Non-Dictatorship: There is no individual i whose preferences always prevail, i.e. there is no $i \in \{1,2, \ldots\ldots, N\}$ such that

$$F\,(R_1, R_2, \ldots\ldots, R_N) = R_i \textit{ for all } (R_1, R_2, \ldots\ldots, R_N) \in L\,(A)^N$$

Independence of irrelevant alternative: For two preference profiles $(R_1, R_2, \ldots\ldots, R_N)$ and $(S_1, S_2, \ldots\ldots, S_N)$ such that for all individuals i, alternatives 'a' and 'b' have the same order in R_i as in S_i, alternatives 'a' and 'b' have the same order

$$F\,(R_1, R_2, \ldots\ldots, R_N) \textit{ as in } F\,(S_1, S_2, \ldots\ldots, S_N)$$

Yu (2012) gives the short proof of Arrow's impossible theorem or we can say Arrow's paradox. He stated the theorem as

"Individual numbered 1,2, ..., N each has complete, reflexive and transitive preference over $M \geq 3$ alternatives $A = \{a_1, a_2, \ldots\ldots, a_M\}$. The set of preference profiles p is unrestricted with a typical element denoted as an ordered list

$\vec{\succcurlyeq} = (\succcurlyeq_1, \succcurlyeq_2, \dots \dots, \succcurlyeq_N)$. A social welfare function *R* assigns to each $\vec{\succcurlyeq}$ complete, reflexive and transitive social preferences $\trianglerighteq$ over A, i.e. *R:* → *P*, where *P* denotes the set of all possible social preferences. Arrow's paradox asserts that all possible social preferences. Arrow's paradox asserts that it is impossible to construct R with the following three properties.

Unanimity: For arbitrary alternative a_i and a_j, if $a_i \succ_n a_j$ (i.e. $a_i \succcurlyeq_n a_j$ and not $a_j \succcurlyeq_n a_i$). for each individual n in $\succcurlyeq$, then $a_i \triangleright a_j$ (i.e. $a_i \trianglerighteq a_j$ and not $a_j \trianglerighteq a_i$).

Arrow's independence of irrelevant alternative (AIIA): If each individual's preferences over a_i and a_j are same in $\vec{\succcurlyeq}$ and $\vec{\succcurlyeq}$,[1] then $R(\vec{\succcurlyeq})$ and $R(\vec{\succcurlyeq}^{\prime})$ rank the two alternatives the same.

Non-dictatorship: There exists no individual n such that for each $\vec{\succcurlyeq} \in$ p and its corresponding $\trianglerighteq = R\,(\vec{\succcurlyeq})$, $a_i \succ_n a_j$ always implies $a_i \trianglerighteq a_j$"

Suppose *R* satisfies unanimity and AIIA. Consider an arbitrate $(\vec{\succcurlyeq})$ in which $a_i \succ_n a_j$ for all n and the swap the position of the two alternatives sequentially from 1to *N*. According to Unanimity, we start with $a_i \triangleright a_j$ and end with $a_j \triangleright a_i$. We call the first voter whose swap invalidates $a_i \triangleright a_j$ the (i, j)- pivotal voter and denote her number n_{ij}. AIIA makes sure that this definition is independent of $(\vec{\succcurlyeq})$.

	$1,\dots\dots,n_{ij}-1$	n_{ij}	$n_{ij}+1,\dots\dots,N$
$\vec{\succcurlyeq}$	$a_j,\dots\dots,a_j$	a_i	$a_i,\dots\dots\dots,a_i$
	$a_k,\dots\dots,a_k$		
	$a_i,\dots\dots,a_i$	a_j	$a_j,\dots\dots\dots,a_j$
		a_k	$a_k,\dots\dots\dots a_k$
$\vec{\succcurlyeq}$	$*,\dots\dots,*$		
	$*a_j,\dots\dots a_i$	a_j	$a_i\dots\dots\dots,a_j$
	$*,\dots\dots,*$		$*,\dots\dots\dots\dots,*$
	$a_i,\dots\dots\dots a_i$	a_i	$*a_j\dots\dots\dots *a_j$
		a_k	$*,\dots\dots\dots\dots,*$

Let us suppose any $\vec{\succcurlyeq}'$ with the depicted rankings of the three alternatives. We must have $a_i \triangleright a_j \triangleright a_K$, where the first relation is by the definition of n_{ij} and the second by Unanimity. For $\vec{\succcurlyeq}$, * denotes possible of a_K, with indifference draw by putting alternatives at same level. We have $a_j \trianglerighteq a_i \trianglerighteq a_K$, where the first in by the definition of n_{ij} and the second by AIIA. For a_j and a_K, we conclude by AIIA that n_{ij} dictates $a_j \triangleright a_K$ i.e.

$$a_j > n_{ij}\, a_k \Rightarrow a_j \triangleright a_k \text{ for all } i \neq j \neq K \tag{1}$$

In the swapping process that defines n_{ij}, equ. (1) says that $a_j \triangleright a_K$ should not change as long asn_{ij} rank jabove *K*, so for n_{ij}, the ranking of two alternatives should become $a_j \triangleright a_K$ no later than n_{ij} make the change, so

$$n_{Kj} \leq n_{ij}$$

we have $\quad n_{jK} \geq n_{ij} \geq n_{ij}$

As *j* and *K* are distinct and arbitrary, $n_{Kj} \geq n_{jK}$ also holds, which gives

$$n_{jK} = n_{Kj} = n_{ij}$$

But equ(1) requires that this unique pivotal voter holds dictatorship over all ordered pair of alternatives, violating Non-dictatorship.

A common way around Arrow's paradox is limiting the alternative set to two alternatives. Thus, whenever more than two alternatives should be put to the test, it seems very tempting to use a mechanism that pairs them and votes by pairs. As tempting as this mechanism seems at first glance, it is generally far from satisfying even Pareto efficiency; not to mention AIIA. The specific order by which the pairs are decided strongly influences the outcome. This is not necessarily a bad feature of the mechanism. Many sports use the tournament mechanism, essentially a pairing mechanism, to choose a winner. This gives considerable opportunity for worker teams to win, thus adding interest and tension throughout the tournament. This means that the person controlling the order by which the choices are paired has great control over the outcome. In any case, when viewing the entire voting process as one game, Arrow's theorem still applies.

Bibliography

1. Arrow, K.T. (1950). A difficulty in the concept of social welfare, Journal of Political Economy, 58(4), pp 328-346.
2. Kirman, A & Sondermann, D. (1972). Arrow's theorem, many agents and invisible dictators, Journal of Economic Theory, 5, pp 267.
3. Mihara, H.R. (1997). Arrow's theorem and Turing computability, Economic Theory, 10(2), pp 257-276.
4. Mihara, H.R. (1999). Arrow's theorem, countable many agents and more visible invisible dictators, Journal of Mathematical Economics, 32, pp 267-277.
5. Suzumura, Kotaro, Arrow, Kenneth, J. and Amartya Kumar (2002). *Handbook of Social choice and welfare*, Vol. 1, Amsterdam, Netherlands, Elsevier, ISBN 978-0-444-829146.
6. Taylor, A.D. (2005). *Social choice and the mathematics of manipulation*, New York, Cambridge University Press, ISBN 0-521-00883-2.
7. Yu, Ning Neli (2012). A one-short proof of Arrow's impossibility theorem. Economic Theory, 50, pp 523-525.

Chapter 7

Benford's Law

NUMBERS STARTING WITH EARLY DIGITS APPEAR DISPROPORTION-ATELY OFTEN IN SEEMINGLY RANDOM DATA SETS.

Benford's law, abbreviated hence forth as BL, is one of the gems of statistical folk lore. It is the observation that in many collections of numbers is they mathematical tables, real-life data, or combinations thereof, the leading significant digits are not uniformly distributed, as might be expected, but are having skewed towards the smaller digits. In general, Benford's law (or first-digit law), refers to the frequency distribution of digits in many real-life sources of data. In the distribution the number 1 occurs as the leading digit about 30% of the time, while larger numbers occur in that positive less frequently 90 as the first digit less than 5% of the time. Ben ford's law also concerns the expected distribution for digits beyond the first which approach a uniform distribution.

The story began in 1881, when the American astronomer Simon Newcomb noticed that books of logarithm tables always seemed grubby on the early pages, and cleans toward the back. For some reason, people seemed to took up numbers beginning with the digits 1 and 2 for more often than they looked up numbers beginning with the digits 8 and 9 actually occurred more often in nature than number beginning with 8 and 9.

Newcomb went so far as to sketch out mathematically the law he expected first digits to follow (Newcomb 1881). A randomly selected number being with digit was 1 about 30% of the time, more precisely, the proportion was 0.301 or $\log_{10}2$. The frequency of numbers with leading digit 2 was about 18%, obtained from $\log_{10}\frac{3}{2}$, those with leading digits 3 should be about 12%, from $\log_{10}\frac{4}{3}$, and so on until the frequency of 8's was 5.1% and that of 9's was 4.6%. Overall, for a number chosen at random from those sought in the logarithm tables of 1881, Newcomb suggested that the probability of the first or leading digit being d should be

$$P\ (leading\ digit = d) = \log_{10}\frac{(d+1)}{d}, \quad d = 1, 2, \ldots\ldots\ldots, 9$$

With an apparent sleight-of-hand reminiscent of Fermat, Newcomb simply stated that his logarithmic rule was 'evident'. It is counter-intuitive enough to think that digits should be anything but uniforms, but this obscure expression for their frequencies has been the last straw for many a baffled on looker.

Ben ford (1938), a physicist with the General Electric Company assembled over 20,000 numbers from sources as diverse as Readers' biggest articles, street addresses of American Men of Science, atomic weights, population sizes, drainage rates of rivers, and physical constants. His data showed that leading digits from a wide range of sources showed an uncanny adherence to the logarithmic rule that Newcomb had panned, apparently unnoticed, decades earlier. Ben fords gave the law its name and a certain mystique, but no convincing explanation. Indeed, he described the phenomenon as belonging to outlaw and anomalous numbers, and reassured us that dapper digits of orderly data such as atomic weights and specific hearts would be free of such absurdities. Ben ford's Law was the province of wild data, population size, addresses of American men of science. Over a third began with digit 1, whereas fewer than one in 20 began with digit 9. According to Ben ford's Law, leading digits as far-fetched as those in the second column would be one-in-a-thousand rarely.

Most of the statisticians and mathematicians come to the conclusion that Ben ford's law is a mysterious law of nature, for accepted that the first rigorous explanation of the law was due to Hill (1995), who provided a measure-theoretical proof that data drawn from a random mix of different distribution rather like Frank Ben ford's original 20,000 will ultimately converge to Ben ford's law.

Many intuitive explanations are given by different statisticians regarding this law such as scale-invariance and base-invariance. But these explanations are not wholly satisfactory because they do not explain why a universal law of nature should arise in the first place. Many authors gives simple explanation using technical language and tools, for example as an application of Poincare's Theorem in circular statistics by Mardia and Jupp (2000), in terms of digital signal processing with Fourier transform by Smith (2007).

Bibliography

1. Benford, F. (1938). The law of anomalous number, Proceeding of the American Philosophical Society, 78, pp 551-572.
2. Durtschi, C., Hillison, W. and Pacini, C. (2004). The Effective use of Benford's law to assist in detecting frond in accounting data, Journal of Forensic Accounting, 5, pp 17-34.
3. Fewster, R.M. (2009), A simple explanation of Benford's law, The American Statistician, 63, pp 26-32.
4. Hill, T.P. (1995). A Statistical Derivation of the Significant Digit Law, Statistical Science, 10, pp 354-365.
5. Mardia, K.V. and Jupp, P.E. (2000). Directional Statistics, Chichester: Wiley, Example 4.1.
6. Newcomb, S. (1881). Note on the Frequency of use of the different digits in national numbers, American Journal of Mathematics, 4, pp 39-40.

7. Raimi, S.W. (2007). The first digit problem, The American Mathematics Monthly, 83, pp 521-538.
8. Smith, S.W. (2007). Explaining Benford Law, Chapter 34 in The Scientist and Engineers Guide to Digital Signal. Processing at http: / / www. dspguide.com

Chapter 8

Berkson's Paradox

IT IS A COMPLICATING FACTOR ARISING IN STATISTICAL TESTS OF PROPORTIONS.

Berkson (1946), a physician- statistician at the Mayo Clinic, published a celebrated paper which cast doubt on the validity of certain types of epidemiologic research carried out in hospital settings. He suggested that the relative frequency of disease in a group of patients who have entered hospitals is inherently biased when compared to whose population served by the hospital. This phenomenon, attributable to the way in which risks of hospitalization combine in patients who have had more than one disease, has become known as Berkson Paradox, or Berkson's Fallacy or Berkson's Bias. The work is of potentially fundamental importance to the epidemiological world. Yet, the possible effect of the bias on hospital based studies that focus on association has been largely ignored in 1930's, presumably due to a lack of supportive evidence for the theory. In the note reported here, an attempt was made to question this apparent complacency by providing an overview of Berkson's Paradox by Roberts et al. (1978).

The idea of verification arose when one of the co-investigators invoked Berkson's arguments at a conference of the National Institutes of Health. Other epidemiologists rebutted by pointing out, quite rightly, that Berkson had only advanced a theoretical objection, never tested. He then consulted with other members of the team about the feasibility of analyzing clinical data, drug use data and health services utilization information which had been gathered by the Mc Master health sciences field survey unit in ongoing primary care research. After encouraging preliminary analysis, they brought together the data from three studies to form the basis of the analysis reported in their study of empirical demonstration of Berkson's bias.

Berkson centered his discussion on the establishment of an association between the exposure to a suspected causal factor and the presence of a disease. In hospital based work, such associations are usually determined through case-control studies. In case control studies the level of exposure to the putative cause is ascertained in a diseased or affected group and in a comparative not affected group. The comparative group would be drawn from patients admitted to hospital with a disease which is not thought to have an association with the suspected causal factor under study. For example, patients with accidental bone fractures might be taken as controls when assessing hypertension as a risk factor for skin cancer because there is no plausible association between such fractures and hypertension.

Mathematically in Berkson's paradox, assume that there are two independent events A and B. These events are not correlated when observed in nature. However, what if one conditions on the fact that either A or B occurs. By conditional on the fact one of the two events occurred, these events are now correlated. If both events have a positive probability less than unity i.e.

$$0 < P(A) < 1 \quad \text{and} \quad 0 < P(B) < 1$$

Then one can describe Berkson's paradox as if $P(A|B)=P(A)$ then $P(A|B,C)<P(A|C) \; where \; C=A\cup B$

The correlation occurs because of selection bias. If some comes out that conditional on C that event A did not occur, we know that event B did occur. Thus, conditioning on the union of the two variables leads to a correlation.

Roberts et al. (1978) showed the three clinical conditions of interest: disease 1, a suspect causal factor thought to be related to the incidence of disease 2

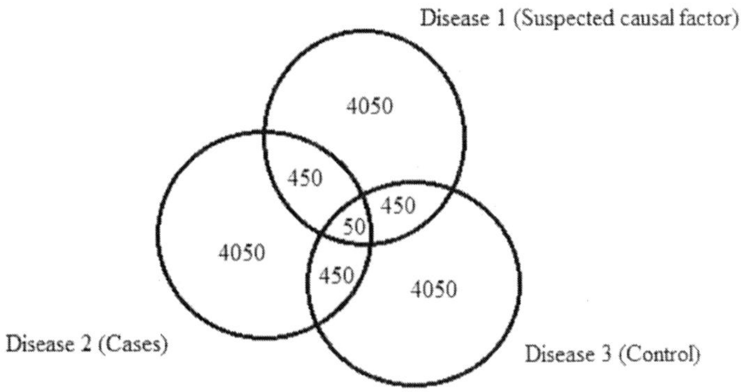

which define cases, and not related to disease,3 which defines controls. In this example, disease 1 would be hypertension, disease 2 skin cancer, and disease 3 fractures. The van diagram depicts the expected distributions of the three diseases as they might arise in a hypothetical free-living population of 50,000 people. It is assumed that the prevalence of each disease is 10% and that they are independent that is the occurrence of one disease in a person does not affect the likelihood of that person also having one of the other diseases. A total of 5,000 people have disease 1, of whom 450 have disease 1 in combination with disease 2 alone, 450 have disease 1 in combination with disease 3 alone and 50 people have disease 1 in combination with both disease 2 and disease 3. They do not find relationship between the suspected calculus 1 factor (disease 1) and the cases with disease 2. This should not be surprising given the starting assumption of independence. The epitome of Berkson's argument rests on the assumption that individual rates of hospitalization for each disease combine together according to the law of probability. The inevitable result is a different net hospitalization rate so that people with multiple diseases or conditions become over represented in the hospital population. He simply discussed the paradox in terms of differences in exposure rates and concluded that the paradox was most apparent when the three selection rates of hospitalization were markedly different. He did not quantify the effect of the bias in terms of relative risk. If the odds ratio estimate of relative risk is used to express the strength of the biased association the algebra simplifies to the point where the expected direction of bias can be described by a simple rule- of- thumb. He pointed out that the bias increase in magnitude with the disparity between hospitalization rates, however only the case and control disease is not a single disease, but 'all other disease' other than the case disease. In this situation the bias is solely determined by the case selection rate and the average hospitalization rate for all other diseases.

We believe that Berkson's bias is an important phenomenon which cannot be overlooked by investigator undertaking epidemiologic or clinical research to show cause and effect etiologic relations in hospital populations.

Bibliography

1. Berkson, J. (1946). Limitation of the application of 4-fold tables to hospital data, Biometrics Bulletin, 2(3), pp 47-53.
2. Cornfield, J. (1951). A method of estimating comparative rates from clinical data: Application to cancer of the lung, breast and cervix, Journal of the National Cancer Institute, 11, pp 1269-1275.
3. Bartlett, M.S. (1935). Contingency table interaction, Journal of Royal Statistical Society (Supplement).,2, pp 248-252
4. Roberts, R.S., Spitzer, W.O., Delmore, T. and Sackett, D.L. (1978). An Empirical Demonstration of Berkson's Bias, Journal of Chronic Diseases, 31, pp 119-128.

Chapter 9

Bertrand's Box Paradox

A PARADOX OF CONDITIONAL PROBABILITY CLOSELY RELATED TO THE BOY OR GIRL PARADOX.

Bertrand's box paradox is a classic paradox of elementary probability theory. It was first posed by Jasept Bertrand in his calculdes probabilitites, published in 1889. According to him three boxes, each with one drawn on each of two sides. Each drawer contains a coin, one box has a gold coin on each side (S_{GG}), one a silver coin on each side (S_{SS}) and the other a gold coin on one side and a sliver coin on other side (S_{GS}). A box is chosen at random, a random drawn is opened, and a gold coin is found inside it. He raised a question "What is the chance of the coin on the other side being gold?" and the following reasoning appears to give a probability of $\frac{1}{2}$

- Originally, all three boxes were equally likely to be chosen.
- The chosen box cannot be boxing (S_{GS}).
- So it must be box (S_{GG}) or (S_{GS})
- The two remaining possibilities are equally likely, so the probability that the box is (S_{GG}) and other coin is also is $\frac{1}{2}$.

The flaw is in the last step. While those two cases were originally equally likely, the fact that you are certain to find a gold coin if you had chosen the S_{GG} box, but only 50% sure of finding a gold coin if you had chosen the S_{GS} box, means they are no longer equally likely given that you have found a gold coin.

- The probability that S_{GG} would produce a gold coin is 1.
- The probability that S_{SS} would produce a gold coin is 0.
- The probability that S_{GS} would produce a gold coin is $\frac{1}{2}$.

Initially S_{GG}, S_{SS} and S_{GS} are equally likely. Therefore by Bayes rule the conditional probability that the chosen box is S_{GG} given that we have found a gold coin is given by

$$\frac{P(Gold|S_{GG})}{P(Gold|S_{GG})+P(Gold|S_{ss})+P(Gold|S_{GS})}=\frac{1}{1+0+\frac{1}{2}}=\frac{2}{3}$$

The correct answer of $\frac{2}{3}$ can also be obtained as follows:

- Originally, all six coins were equally likely to be chosen.
- The chosen coin cannot be from drawer S of box S_{GS} or from either drawer of box S_{SS}.
- So it must come from the G drawer of box S_{GS} or either drawer of box S_{GG}.
- The three remaining possibilities are equally likely, so the probability that the drawer is from box S_{GG} is $\frac{2}{3}$.

Alternatively, one can simply not that the chosen box has two coins of the same type $\frac{2}{3}$ of the time. So, regardless of what of coin is in the chosen drawer, the box has two coins of that type $\frac{2}{3}$ of the time. In other words, the problem is equivalent to asking the question "What is the probability that 1 will pick a box with two coins of the same color?" Bertrand's point in constructing this example was to show merely counting cases is not always proper. Instead, one should sum the probabilities that the cases would produce the observed result and the two methods are equivalent only if this probability is either 1 or 0 in every case.

Bertrand originally described it after a box has been chosen, but before a box is opened to let you observe a coin, the probability is $\frac{2}{3}$ that the box has two of the same kind of coin. If the probability of observing a gold coin in combination with the box has two of the same kind of coin is $\frac{1}{2}$, then the probability of observing a sliver coin in combination with the box has two of the same kind of coin must also be $\frac{1}{2}$ and if the probability that the box has two like coins changes to $\frac{1}{2}$ no matter what kind of coin is shown, the probability would have to be $\frac{1}{2}$ even if you hadn't observed a coin this way. Since we know his probability is $\frac{2}{3}$, not $\frac{1}{2}$, we have an apparent paradox. It can be resolved only by recognizing how the combination of observing a gold coin with each possible box can only affect the probability that the box was gold coin on each side (S_{GG}), silver coin on each side (S_{SS}) and gold coin on one side and a sliver coin on other side (S_{GS}).

Bibliography

1. Dale, A.I. (1999). *A history of inverse probability: From Thomas Bayes to Karl Pearson*, Springer, New York.
2. Bertrand, J. (1889). Calcul des probabilities, Gauthier-Villars, Paris.
3. Gelman, A. Carlin, J.B., Stem, H.S. and Rubin D.B. (2003). *Bayesian Data Analysis*, Second Ed. CRC Press.
4. Stone, J.V. (2013). Bayes Rule: A Tutorial Introduction University of Sheffield, Psychology.

Chapter 10

Bertrand's Paradox

DIFFERENT COMMON-SENSE DEFINITIONS OF RANDOMENESS GIVE QUITE DIFFERENT RESULTS.

In the theory of geometrical probabilities the random elements are not quantities but geometrical objects such as points, lines and rotations. Since the ascription of a measure to such elements is not quite an obvious procedure, a number of paradoxes can be produced by failure to distinguish the reference set. In 1889, Bertrand published a collection of paradoxes the most celebrated 0 which is his "random-chord" problem. This particular problem, often labeled simply "Bertrand's Paradox".

Bertrand (1889) gave an example to show the probabilities may not be well defined if the mechanism or method that produces the random variable is not clearly defined. What is the probability that the chord is longer than a side of the inscribed equilateral triangle of the circle? According to him, it is a paradox because the issue has at least three solutions that contradict to each other but each is derived from seemingly decent premises and reasoning.

The first solution is, a chord is longer if the angle between the chord and the tangent line passing an endpoint of the chord on the circle is within the range $60^0 \sim 120^0$. Therefore, the probability is $\frac{1}{3}$. The second solution is the chord is longer if its midpoints lies within a concentric circle with half the original radius, so, since the area of this inner circle is a quarter that of the original circle, the probability is $\frac{1}{4}$. The third solution is the chord is longer if its midpoint lies on the inner half of the radius bisecting it, so since the midpoint may lie anywhere on this radius, the probability is $\frac{1}{2}$.

These are three solutions given in Bertrand's paradox, actually derived by using different ways of generating random chords.

Method 1: 1-point-on-circumference, 1-angle

Pick a point X on the circumference of the given circle randomly, which defines one intersection of the chord with the circle. Randomly pick a number between 0 and 180 which is the angle between the chord to be generated and the tangent line of the circle at X.

Method 2: 1-internal-point as chord-middle-point

Pick up a point inside the given circle randomly, which defines a chord taking the point as the middle- point of the chord to be generated.

Method 3: 1-point-on-diameter as chord-middle point

Randomly pick a diameter line in the given circle, by randomly selecting the angle of the diameter line. Pick a point randomly on the diameter line, which defined a chord by taking the point as the middle- point the chord to be generated.

The three methods have been viewed indifferent for the purposes of generating random chords in the past two hundred years, and thus perplexed people with the different solutions derived from these methods.

Analysis of method of generating random chords

We evaluate the three random chord generation method by showing whether the lines or chords are generated purely randomly without any dubious restrictions and/or whether the generated chords violates the propositions or corollaries of characterization of uniform distribution of lines.Wang(2008) discussed the solutions which are provided here.

Lines or chords generated at random without any additional conditions are purely random and therefore uniformly distributed. "Tossing broom straws" is an example of generating purely random lines. Chords at random as mentioned initially by Bertrand (1889) should refer to the chords with pure randomness.

The method 1 is the one that leads to solution 1. This method is a way equivalent to randomly generating two points on circumference. They share the same assumption that the end points of random chords are uniformly distributed along the circumference, which is false according to well-known corollary, a set of parallel lines in a finite set of uniformly distributed straight lines in two dimensional space do not have constant density along a curve on which directions change. The fallacy of this method can also be seen in this way. As the result of this method, the end points of all generated chords are uniformly distributed along the circumference; therefore, the end points of any set of parallel chords are uniformly distributed lines. Therefore, the chord generated by this method is not uniformly distributed and not valid random chord.

Method 2 leads to solution 2. The assumption of this method is that middle points are uniformly distributed over the area inside a circle. But this assumption is obviously wrong. The point at the centre of the circle has more chance to be a middle point than a point near circumference which can only be middle point for the chords perpendicular to the diameter through it. So, the well known proportion, every point in two dimensional space has an equal chance to be on a straight line is violated. The lines thus generated are not uniformly distributed.

Method 3 leads to solution 3. This method follows the following proposition of uniformly distributed lines.

Let *t* be a straight line in two dimensional space *S*. Let *U* denote a sub-set of a finite set of uniformly distributed straight lines in *S*, says *B*, so that it contains parallel lines in a direction that is not parallel to *t*. Let us define $D_{on\text{-}t\text{-}by\text{-}U}$ as density of intersection points of lines in *U* on line t so the $D_{on\text{-}t\text{-}by\text{-}U}$ = number of intersection points on t per unit length, then

Proposition: $D_{on\text{-}t\text{-}by\text{-}U}$ is expected to be a constant along the line t.

Proposition: Let *s* and *t* denote two non-parallel straight lines.

Let $U \subset B$ denote the set of parallel lines perpendicular to *s*, and $V \subset B$ the set of parallel lines perpendicular to *t*. Then $D_{on\text{-}t\text{-}by\text{-}U} = D_{on\text{-}t\text{-}by\text{-}V}$

As far as the diameter is generated in purely random way the chords thus are valid "random chord and uniformly distributed. To generate a diameter, one may randomly choose an angle of the circle center, which is equivalent to picking up the diameters of the circle from the uniformly distributed lines.

Jayne's (1957) mathematically developed three features for the random chords in Bertrand's paradox: rotational invariance, scale invariance, and translational invariance, which led to the conclusion that the correct answer to Bertrand's paradox is $\frac{1}{2}$. What Jayne's did not do was proving the other answers are wrong. Therefore the myth of multiple solutions of Bertrand's paradox has remained up to the 1973. But Wang (2008) showed the solution 1 and solution 2 for Bertrand's paradoxes are not valid, and only solution 3, $\frac{1}{2}$, is the correct answer.

Bibliography

1. Bertrand, J. (1889). Calcul des probabilities, Gauthier-Villars, Paris, pp 4-5.
2. Guiasu, S. and Shenitzer, A. (1985). The principle of maximum entropy, The Methematical Intelligencer, 7(1),
3. Joynes, E.T. (1957). Information Theory and Statistical Mechanics, Physical Review, 106(4), pp 620-630.
4. Wang, J. (2008). Bertrand's paradox and distribution of lines, proceeding of DSI2008, Annual National Conference, Baltimore, pp 3031-3036.

Chapter 11

Birthday Paradox

WHAT IS THE CHANCE THAT TWO PEOPLE IN A ROOM HAVE THE SAME BIRTHDAY?

In probability theory, the birthday paradox concerns the probability that, in a set of *n* randomly chosen people, some pair of them will have the same birthday. This is not a paradox in the sense of leading to a logical contradiction, but is called a paradox because the mathematical truth contradicts naïve intuition. An intuitive guess would suggest that the chance of two individuals sharing the same birthday in a group 23 is much lower than 50%, but the birthday problem demonstrates that this is more than 50%. According to pigeonhole principle, which state that if n items are put into m pigeonhole with $n > m$, then at least one pigeonhole must contain more than one item; the probability reaches 100% when the number of people reaches 367. However, 99% probability is reached with just 57 people and 50% probability with 23 people. These conclusions include the assumption that each day of the year, except Feb. 29, is equally probable for a birthday. The history of this problem is obscure, but Ball (1960) indicated that it was first discussed by an H. Davenport. The probabilistic reasoning behind this problem led to well-known cryptographic attack called the birthday attack which can be used to abuse communication between two or more parties. The attack depends on the higher likelihood of collisions found between random attack attempts and a fixed degree of permutations.

The Birthday paradox is to find the probability that, in a group of *n* people, there is at least one pair of people who have the same birthday. In the list of 23 people, comparing the birthday of the first person on the list to the others allows 22 chances for a matching birthday, the second person on the list to the others allows 21 chances for a matching birthday, and third person has 20 chances, and so on. Hence

$$\textit{Total Chances} = 22 + 21 + 20 + \ldots\ldots\ldots\ldots + 1 = 253$$

so comparing every person to all of the others allows 253 distinct chance i.e. in a group of 23 people there are

$$\binom{23}{2} = \frac{23 \times 22}{2} = 253\,pairs$$

Presuming all birthdays are equally probable, the probability of a given birthday for a person chosen from the entire population at random is $\frac{1}{365}$. Although the pairing in a group of 23 people are not statistically equivalent to 253 pairs chosen independently, the birthday paradox becomes less surprising if a group is thought of in terms of the number of possible pair, rather than as the number of individuals.

Here, the problem is to compute the approximate probability that in a room of n people, at least two have the same birthday (for ingenuity, disregard variations in the distribution, such as leap years, twin's seasonal or weekly variations and assume that the 365 possible birthday are equally likely). Real life birthday distribution is not uniform since not all dates are equally likely. If P *(SB)* is the probability of at least two people in the room having same birthday, it may be simpler to calculate P *(SB')*, the probability of there not being any two people having the same birthday, then people is 1, or 100%. The probability of 1 can also be written as $\frac{365}{365}$ as it is

$$P(SB) = 1 - P(SB')$$

because *SB* and *SB'* are the only two possibilities and are also mutually exclusive cases.

When events are independent of each other, the probability of all of the events occurring is equal to a product of the probabilities of each of the events occurring. Therefore, if *P (SB')* can be described as 23 independent events, P*(SB')* be calculated as

$$P(1) \times P(2) \times P(3) \times \ldots\ldots\ldots\ldots\ldots \times P(23)$$

The 23 independent events correspond to the 23 people and can be defined in order. Each event can be defined as the corresponding person not sharing his/her birthday with any of the previously analyzed people. For event 1, there are no previously analyzed people. Therefore the probability *P*(1), that person number 1 does not share his/her birthday with previously analyzed equally likely to happen on each of the 365 days of the year. The probability *P*(2), that person 2 has a different birthday than person 1 is $\frac{364}{365}$. Because, if person 2 was born on any of the other 364 days of the year, person 1 and 2 will not share the same birthday. Similarly, if person 3 is born on any of the 363 days of the year other than the birthday of person 1and 2, person 3 will not share their birthday. This makes the probability $P(3) = \frac{363}{365}$. This analysis continues until person 23 is reached, which is equal to $\frac{343}{365}$, therefore *P (SB')* is equal to the product of these individual probabilities

$$P(SB') = \frac{365}{365} \times \frac{364}{365} \times \frac{363}{365} \times \ldots\ldots\ldots\ldots\ldots \times \frac{343}{365}$$

$$= \frac{1}{(365)^{23}}[365 \times 364 \times 363 \times \ldots\ldots.343]$$

$$\approx 0.492702766 \approx 49.27\%$$

Thus $$P(SB) = 1 - P(SB')$$

$$= 0.507297 \approx 50.73\%$$

This process can be generalized to a group of n people. Let us consider *p (n)* is the probability of at least two of the n people sharing a birthday. Then, according to pigeonhole principle $\bar{p}$ (n) is given by

$$\bar{P}(n) = 1 \times \left(1 - \frac{1}{365}\right) \times \left(1 - \frac{2}{365}\right) \times \ldots\ldots\ldots \times \left(1 - \frac{n-1}{365}\right)$$

$$= \frac{365 \times 364 \times \ldots\ldots\ldots \times (365 - (n-1))}{365^n}$$

$$= \frac{365!}{365^n (365 - n)!} = \frac{n! \binom{365}{n}}{365^n}$$

Some statistician proves the same result in another way. Firstly they define *S* to be a set of *N* people and *B* be the set of dates in a year. They define the birthday function

$$b: S \to B$$

to be the map that sends a person to their birthday. Then it is obvious that everyone in *S* has a unique birthday if and only if the birthday function is injective. They considered how many functions, and how many injective function, exist between Sand *B*,

Since $|S| = N$ and $|B| = 365$

it follows that there are 365^N possible function and

$$\frac{365!}{(365 - N)!}$$

is the possible injective function.

Let A be the statement "Everybody in *S* has a unique birthday", the by definition

$$P(A') = \frac{365!}{365^N (365 - N)!} \qquad or\ P(A)$$

therefore

$$P(A) = 1 - \frac{365!}{365^N (365 - N)!} \qquad or\ P(A')$$

Bibliography

1. Abramson, M. and Moser, W.O.J. (1970). More Birthday Surprises, American Mathematical Monthly, 77, pp 856-858.
2. Ball, W.W. Rouse (1960). Other *Question on Probability in Mathematical Recreations and Essays,* Macmillan, New York, pp 45.
3. Borgs, C., Chayes, J. and Pitter, B. (2001). Phase Transitio and Finite Size Scaling in the Integer Partition Problem, Random Structures and Algorithm, 19, pp 2470288-247-288
4. Brink, D. (2012). A exact solution to the Birthday Problem, Ramanujan Journal, 28 (2), pp 223-238
5. Mathers, C.D. and Harris, R.S. (1983). Seasonal Distribution of Births in Australia, International Journal of Epidemiology, 12(3), pp 326-331.
6. Mathis, Frank, H. (1991). A Generalized Birthday Problem, Society for Industrial and Applied Mathematics Review, 33(2), pp 265-270.
7. Wendi, M.C. (2003). Collision probability between set of random variable, Statistics and Probability Letters, 64(3), pp 249-254.

Chapter 12

Borel - Kolmogorov Paradox

CONDITIONAL PROBABILITY DENSITY FUNCTIONS ARE NOT INVARIANT UNDER COORDINATE TRANSFORMATIONS.

In probability theory, the Borel - Kolmogorov paradox, sometimes known as Borel Paradox, is a paradox relating to conditional probability with respect to an event of probability zero. It is named after Emile Borel and Andrey Kolmogorov (1933). To understand Borel paradox we have to choose basic set *E*, the set of all points on a spherical surface and *P(A)* is to be proportional to the measure of set *A*. Let us now choose two diametrically opposite points for our poles, so that each meridian circle will be uniquely defined by the longitude ψ, $0 \leq \psi < \pi$. Since ψ varies from 0 to π, in other words, we are considering complete median circle, the latitude θ must vary from $-\pi$ to $+\pi$. The problem is to determine "the conditional probability distribution of latitude θ, $-\pi \leq \theta < +\pi$ for given longitude ψ. This way to explain this problem is quite difficult. Many statisticians explained this problem in different ways. Jayne's (2003) explain this problem using the Bayesian probability theory and is considered here to explain this paradox using this theory because it easy and simple to understand.

Suppose *I* am prior information according to which *(x,y)* are assigned a bivariat normal pdf with variance unity and correlation coefficient *p*.

$$P(dxdy|I) = \frac{\sqrt{1-\rho^2}}{2\pi} \exp\left[-\frac{1}{2}\left(x^2 + y^2 - 2\rho xy\right)\right] dxdy \quad (1)$$

We can integrate out either x or y to obtain the marginal pdf's.

$$P(dxdy|I) = \left(\frac{1-\rho^2}{2\pi}\right)^2 \exp\left[-\frac{1}{2}\left(1-\rho^2\right)x^2\right] dx \quad (2)$$

$$P(dy|I) = \left(\frac{1-\rho^2}{2\pi}\right)^2 \exp\left[-\frac{1}{2}\left(1-\rho^2\right)y^2\right] dy \quad (3)$$

Now the question is "What is the conditional pdf for x, given that y = y_0?" for this we need to take y = y_0 in equation1 and renormalize as

$$P(d\,x|y = y_0, I) = A \exp\left[-\frac{1}{2}\left(x^2 + y_0^2 - 2\rho x y_0\right)\right] dx \quad (4)$$

(where*A* is a normalizing constant, But there is no guarantee that this is valid, because we have obtained by an intuitive ad hoc. Device we did not derive it by applying the basic rules of probability theory). For discrete case

$$P(AB|x) = p(AB|x) = p(A|Bx)\,p(B|x)$$

for which a discrete conditional probability is given by the

$$P(A|B\,x) = \frac{p(AB|x)}{P(B|x)}$$

often taken as the definition of a conditional probability. But we can do the calculation by strict application of our rules if we define the discrete propositions

$$A \equiv x \text{ in } dx$$

$$B \equiv y \text{ in } y_0 < y < y_0 + dy$$

then we should write instead of equation (4), using equation (2) and equation (3).

$$P(A|BI) = p(d\,x|dyI) = \frac{p(dxdy|I)}{P(dy|I)}$$

$$= \frac{1}{\sqrt{2\pi}} \exp\left[-\frac{1}{2}(x - py_0)^2\right] dx \tag{5}$$

Since dy cancels out, taking the limit dy → 0 does nothing Now we find that equation (4) and equation (5) are almost identical. The reasoning that led us to equation (4) happened to give a correct result here, but it can equally well yield any result we please instead equation (4). It depends on the particular form in which we choose to write our equations and therefore generate a paradox. To show this suppose that we had used instead of (x, y) the variables (x, u) where

$$u = \frac{y}{f(x)} \tag{6}$$

with $0 < f(x) < \infty$. The Jacobian is

$$\frac{\partial(x,\mu)}{\partial(x,y)} = \left(\frac{\partial\mu}{\partial x}\right)_x = \frac{1}{f(x)}$$

So the pdf given in equation (1) can be expressed in the new variables

$$P(dxdu|I) = \frac{\sqrt{1-\rho^2}}{2\pi} \exp\left[-\frac{1}{2}\left(x^2 + u^2 f^2(x) - 2\rho u f(x)\right)\right] f(x)dxdu$$

Again, we can integrate out uor x, leading to a marginal distribution $p(du|I)$ which is easily seen to be identical with equation (2) and $p(du|I)$ which is found to be identical with equation (3), transformed to the variable u. Then the conditional pdf for x, given that $u = 0$ (following the reasoning that led us to equation (4)), we find

$$P(dx|u = 0,1) = A\exp\left(-\frac{1}{2}x^2\right) f(x)dx \tag{7}$$

Now from equation (6), the conditional $u = 0$ is the same as $y = 0$, so it appears that this should be the same as equation (4) with $y_0 = 0$. But equation (7) differs from that by an extra factor f(x) which could be arbitrary. Many find this astonishing and unbelievable, but the condition u = 0 is exactly the same condition as y = 0, how can there be a different result?

In the present problem our discrete hypotheses were of the form $a \leq y \leq b$, placing y in an interval of positive measure b–a. Then by the proposition y=0, which has measure zero,we could mean only the limit of some sequence of propositions referring to positive measure, such as

$$A_\in \equiv |y| < \in$$

as $\in \to 0$. The propositions $A_\in$ confined the point (x, y) to successively narrower horizontal stripes, but for any $\in > 0$, $A_\in$ is a discrete proposition with a definite positive probability, so by the product rule the conditional rule the conditional probability of hypothesis $H \equiv x \in dx$

$$p(H|A_\in I) = \frac{p(H, A_\in |I)}{p(A_\in |I)} \tag{8}$$

is well defined and the limit of this as $p(C_i|G_j) = \frac{p(C_i \text{ and } G_j)}{p(G_j)}$ is also a well defined quantity. But the proposition y = 0 may defined equally well as the limit of the sequence.

$$B_\in \equiv |y| < \in |x|$$

of successively thinner wedges and $p(H|B_\in I)$ is also unambiguously defined in equation (8) for all ε > 0. Yet although the sequence $\{A_\in\}$ and $\{B_\in\}$ tend to the same limit y = 0, the conditional densities tend to different limits

$$\lim p(H|A_\in) \alpha\, g(x)$$

$$\lim p(H|A_\in) \alpha\, |x| g(x)$$

and in place of |x| we could put an arbitrary non-negative function $f(x)$. Merely to specify $y = 0$ without any qualifications is ambiguous, it tells us to pass to a measure-zero limit, but does not tell us which of any number of limits is intended. Sometimes we have a probability density on one space and we wish to generate from it one on a subspace of measure zero, the only safe procedure is to pass to an explicitly defined limit by a process like, equation (8).

In general, the final result will depend on which limiting operation was specified. This is extremely counter- intuitive at first hearing, yet it becomes obvious when the reason for it is understood.

Bibliography

1. Jaynes, E.T. (2003). Probability theory: The Logic of Science, Cambridge University Press, pp 467-470.
2. Kolmogorov, Andrey (1933). *Grundbegriffe der Wahrscheinlichkeitsrechnung (in German)*, Berlin, Julius, Springer.
3. Kolmogrorov, Andrey (1956), Translation of above, Foundations of the Theory of Probability (2nd ed.), New York, Chelsea, pp 50-51.
4. Mosegaard, K. and Tarantola, A. (2002) 16 Probabilistic approach to inverse problem, International Geophysics, 81, pp 237-265.
5. Pollard, D. (2002). A User's Guide to Measure Theoretic Probability, Cambridge University Press, pp 122-123.

Chapter 13

Boy or Girl Paradox

A TWO-CHILD FAMILY HAS AT LEAST ONE BOY.
WHAT IS THE PROBABILITY THAT IT HAS A GIRL?

For at least a century, probability devotees have puzzled over a problem known as the Boy or Girl paradox. The Boy or Girl paradox surrounds a well-known set of questions in probability theory which is also known as **"The Two Child Problem"**. The initial formulation of the question dates back to at 1959, when Martin Gardner published on of the earliest variants of the paradox in Scientific American, titled The Two Children Problem. In his study he raise the question what is the probability that there are two boys in the family of two children. There are no further conditions. There are four possible family configurations $\Omega = \{BB, BG, GB, GG\}$ where, for example, BG signifies that the first-born child is a boy and the second a girl. We arrange these in any array

$$\begin{bmatrix} BB & BG \\ GB & GB \end{bmatrix} \qquad (1)$$

As each of the four possibilities is equally likely, we can assign equal relative frequencies or weights to all. In particular, two boys occur with probability $p = \frac{1}{4}$.

Now consider the second problem: the first-born child is a boy. The sample space in now Ω = {BB,BG}, i.e., we consider only the first row of (1). As both events are equally likely, the probability of two boys is $p = \frac{1}{2}$. For the third problem, where the condition is that at least one of the children is a boy, the sample space is $\Omega = \{BB, BG, GB\}$ we retain the first row and the first column of (1). As each event is equally likely, the probability of two boys is $p = \frac{1}{3}$. This is the problem that Martin Gardner popularized, it is often called the Two-Child Paradox. The intuitive answer is $p = \frac{1}{2}$, whereas mathematical reasoning leads us to the answer $p = \frac{1}{3}$.

However, if the family was first selected and then a random, true statement was about the gender of one child, the correct way to calculate the conditional probability is not count the cases that match. Instead, one must add the probabilities that the conditions will be satisfied in each case

Older Child	Younger Child	Probability	P(A)	P(B)
Girl	Girl	$\frac{1}{4}$	0	0
Girl	Boy	$\frac{1}{4}$	$\frac{1}{2}$	$\frac{1}{8}$
Boy	Girl	$\frac{1}{4}$	$\frac{1}{2}$	$\frac{1}{8}$
Boy	Boy	$\frac{1}{4}$	1	$\frac{1}{4}$

where A is the event that at least one boy (given this case)

B is the event that at least one boy (both this case)

The answer is found by adding the number in the last

column wherever we would have counted that case

$$\frac{\frac{1}{4}}{0+\frac{1}{8}+\frac{1}{8}+\frac{1}{4}}=\frac{1}{2}$$

It should be note that this is not necessarily the same as reporting the gender of a specific child, although doing so will produce the same result by a different calculation. For instance, if the younger child is picked, the calculation is

$$\frac{\frac{1}{4}}{0+\frac{1}{4}+\frac{1}{4}+0}=\frac{1}{2}$$

In general, $\frac{1}{2}$ is a better answer any time with a boy and a girl could have been identified as having at least one girl.

On the other hand, there is another approach, the Bayesian approach, it begins with prior probabilities of one-fourth for each of the four possibilities i.e.,

$$P(BG) = P(GB) = P(GG) = P(BB) = \frac{1}{4}$$

It then uses Bayes Theorem to calculate "posterior probabilities" that are conditional on the new information about family. One attractive feature of the Bayesian approach is that it encourages us to think about how the information was obtained. Suppose that we learn the gender of one of the children because the mother mentions him or her. In this case, the mother mentions a girl, G_m and we wish to find the conditional probability that the family is *GG*, given this information, $P(GG \mid G_m)$

Applying Bayes Theorem, which utilizes the multiplication and addition rules of probability, we get

$$P(GG|G_m)=\frac{P(GG \, and \, G_m)}{P(G_m)}$$

$$= \frac{P(GG\, and\, G_m)}{P(GG\, and\, G_m) + P(BG\, and\, G_m) + P(GB\, and\, G_m) + P(BB\, and\, G_m)}$$

$$= \frac{P(G_m|GG) \times P(GG)}{P(G_m|GG) \times P(GG) + P(G_m|BG) \times P(BG) + P(G_m|GB) \times P(GB) + P(G_m|BB) \times P(BB)}$$

If the mother has two daughters, she can only mention a girl, $P(G_m | GG) = 1$. If she has two sons, she can only mention a boy, $P(G_m | BB) = 0$. If she has a son and a daughter, a plausible assumption is that she is equally likely to mention either child.

$$P(G_m | BG) = P(G_m | GB) = \frac{1}{2}$$

Plugging these conditional probabilities and the prior probabilities, into the formula, we get

$$P(GG | G_m) = \frac{1}{2}$$

The calculation can be done succinctly using a contingency table and we will use that approach in the rest of our presentation. For the case just examined, following table shows the expected number of families for various contingencies based on a random sample of 400 families.

	BB	*BG*	*GB*	*GG*	*Total*
Mentions Boy	100	50	50	0	200
Mentions Girl	0	50	50	100	200
Total	100	100	100	100	400

The entries in the interior of the table equal joint probabilities like $P(GG$ and $G_m)$ multiplied by 400, while the totals equal marginal probabilities like $P(G_m)$, for one of the row or $P(GG)$, for one of the column multiplied by 400. For example, we expect there to be 100 *BG* families and that in half of these the mother would mention the boy and in the other half the girl. Notice that the outcomes *BG*, *GB* and *GG* are not equally likely once the mother mentions the gender of one of the children, contrary to the assumption of the sample space approach. One consider the joint probability *P(GG* and G_m) equals $P(G_m|GG)=\frac{1}{4}$ This implies an expected number of families equal to 100. It is twice the joint probability of the family being *BG* and the girl being mentioned, for which the expected number of families is 50. However, no matter whether the mother mentions a boy or a girl, her other child is equally likely to be a boy or a girl. If she mentions a girl, we can use the entries from the relevant row of the table to see that the probability that the other child is a boy is given by $\left(\frac{50+50}{200}\right)=\frac{1}{2}$, while the probability that the other child is a girl is given by $\frac{100}{200} = \frac{1}{2}$.

One third would be the correct answer only "under the assumption that the mother never mentions a son if she has a daughter, $P(G_m|BG)=P(G_m|GB)=1$. Based on the expected numbers of families shown in the following table, which applies the assumption, the probability of a two-girl family equals one-third if the mother mentions a girl and the probability of a two-boy family equal one if she mention a boy. This extreme assumption is never included in the presentation of the two-child problem and is surely not what people have in mind when they present it.

	BB	BG	GB	GG	Total
Mentions Boy	100	50	50	0	200
Mentions Girl	0	50	50	100	200
Total	100	100	100	100	400

A general question is how best to accommodate new information into the evaluation of uncertain situations. Use of the restricted sample space approach for the two-child problem does not yield a proper conditional probability that a family has, two girls, given that one has learned that one of the children is a girl. All it offers in this case is a hypothetical calculation of the fraction of BG, GB and GG families that are GG. In the classic two-child problem, it also offers an erroneous illusion of simplicity, that in general a two-child family is equally likely to be BG, GB or GG if we learn that one of the children is a girl.

In contrast, the Bayesian approaches provide useful conditional probabilities that can be applied directly to a family at hand. It also provides a discipline in that it requires us to be clear about full set of assumptions that enter into our probabilistic inferences. We believe that we have made plausible assumptions about the probability that a mother would mention a girl versus a boy, depending on whether the family is BB, BG, GB or GG, but the Bayesian approach offers the flexibility to accommodate alternative sets of assumptions in any case.

Bibliography

1. Bar-Hillel, M. and Falk, R. (1982). Some teasers concerning conditional probabilities, Cognition, 11, pp 109-122.
2. Freund, J.E. (1965). Puzzle or paradox?, The American Statistician , 19, pp 29-44.
3. Gardner, M. (1959). The Scientific American Book of Mathematical Puzzles and Diversions, New York, Simon and Schuster.
4. Nickerson, R.S. (1996). Ambiguities and unstated assumptions in probabilistic reasoning, Psychological Bulletin, 120,pp 410-433.

Chapter 14

Buridan's Ass Paradox

HOW CAN A RATIONAL CHOICE BE MADE BETWEEN TWO OUTCOMES OF EQUAL VALUE?

Buridan's ass is an illustration of a paradox in philosophy, it refers to a hypothetical situation wherein an ass that is equally hungry and thirsty is placed precisely midway between a stack of hay and a pail of water. Since the paradox assumes the ass will always go to whichever is closer, it will die of both hunger and thirst since it cannot make any rational decision to choose one over the other. The paradox is named after the 14th century French philosopher Jean Buridan, whose philosophy of moral determinism it satirizes. A common variant of the paradox substitute's two identical piles of hay for the hay and water, the ass, unable to choose between the two, dies of hunger. Some proponents of hard determinism have granted the unpleasantness of scenario, but have denied that it illustrates a true paradox, since once does not contradict oneself, is suggesting that a man might die between two equally plausible routes of action. But some writers have been opted to deny the validity of the illustration. A typical counter- argument is that rationally as described in the paradox is so limited as to be a straw man version of the real thing, which does allow the consideration of meta-arguments. In other words, it is entirely rational to recognize that both choices are equally good and arbitrarily pick one instead of starving. This counter-argument is sometimes used as an attempted justification for faith or intuitively. The argument is that, like the starving ass, we must make a choice in order to avoid being frozen in endless doubt.

The situation of Buridan's ass was given a mathematical basis in a 1984 paper by American computer scientist Leslie Lamport, in which Lamport present an argument that, given certain assumptions about continuity in a simple mathematical model of the Buridan's ass paradox, there will always be some starting condition under which the ass will starve to death, no matter what strategy it takes. With the benefit of modern mathematics, the argument can be illustrated as. Assume that, at time zero (0), the ass is placed at position x along the line joining the bales of hay, where one bale is at position 0 and the other at position 1, so $0 < x < 1$. The position of the ass at time $t > 0$ is a function of two arguments, the time t and the starting position x. Let us assume that $A_t(x)$ denote that position. Also assume that when the ass reaches a bale of hay it stays there forever so for all $t \geq 0$,

$$A_t(0) = 0$$

And $$A_t(1) = 1 \qquad 0 \leq A_t(x) \leq 1$$

$$0 < x < 1$$

The ass is a physical mechanism subject to the laws of physics. Any such mechanism is continuous so $A_t(x)$ is a continuous function of x. Since $A_t(0) = 0$ and $A_t(1) = 1$, by continuity there must be a finite range of values of x for which $0 \leq A_t(x) \leq 1$. These values represent initial positions of the ass for which it does not reach either bale of hay within t seconds. Such a range of values of x exists for any time t, including times large enough to insure that the ass has been starved to death by them. Thus, there exists a finite range of starting positions for which the ass starves to death. The key assumption in this argument is continuity, the ass's position at a later time is a continuous function of its initial position.

The general principle underlying the starvation of Buridan's ass paradox can be stated as

> "A discrete decision based upon an input having a continuous range of values cannot be made within a bounded length of time."

Buridan's ass starves because it cannot make the discrete decision of which pile of hay to eat, a decision based upon an initial position having a continuous range of values, within the bounded length of time before it starves. A continuous mechanism must either forgo discreteness, permitting a continuous range of decisions or must allow an unbounded length of time to make the decision. Buridan's principle rests upon mathematical concepts of continuity and boundedness that are not physically observable. No real experiment, having finite precision, can demonstrate the presence or absence of continuity, which is defined in terms of limits. No experiment can demonstrate that an arbiter requires an unbounded length of time it reaches decision.

To understand the meaning of Buridan's principle as a scientific law, consider the analogous problem with classical mechanics. Kepler's first law states that the orbit of a planet's is an ellipse. This is not experimentally verifiable because any finite-precision measurement of the orbit is consistent with an infinite number of mathematical curves. Similarly, we cannot conclude from Buridan's principle that the decision will take an unbounded time what we can conclude is that it may take a very long time to make the decision. The significance of Buridan's principle lies in its warning that decisions may, in rare circumstances, take much longer than expected. Real accidents may occur because people cannot decide in time which of two alternative actions to take even though either would prevent the accident. Although Buridan's principle implies that the possibility of such an accident cannot be eliminated, awareness of the problem could lead to methods for reducing its probability Buridan's principle appears to be a universal law of nature, perhaps a fundamental law not derivable from others.

Modern interest in Buridan's principle stems from the arbiter problem, its manifestation in computers. There is a long history of computer scientists being unaware of the problem and even denying its existence. Also a version of this principle occurs in electrical engineering these days. Specifically, the input into a digital logic gate much convert a continuous voltage value into a 0 or 1 which is typically sampled and then processed. Here the voltage value can then be likened to the position of the ass and the values 0 and 1 represent the bales of hay.

Bibliography

1. Kane, Robert (2005). *A Contemporary Introduction to Free Will,* New York, Oxford, pp 37.
2. Kinniment, D.J. (2008). *Synchronization and Arbitration in Digital System,* John Wiley & Sons, pp 3, ISBN 0470517131.
3. Mawson, T.L. (2005). *Belief in God,* New York, NY: Oxford University Press, pp 201.
4. Rescher, Nicholas (1950/60). Choice Without Preference: A Study of the History and of the Logic of the Problem of Buridan's Ass, Kant- Studien, 51, pp 142-175.
5. Rescher, Nicholas (2005). *Cosmos and Logos: Studies in Greek Philosophy,* Ontos Verlag, pp 93-99, ISBN 393720265X.
6. Ullmann-Margalit, E. and Morgenbesser, S. (1977). Picking and Choosing, Social Research, 44, pp 757-785.
7. Zupko, Jack (2003). *John Buridan: Portrait of Fourteenth Century Arts Master,* Notre Dame, Indiana: University of Notre Dame Press, pp 258.

Chapter 15

Chain-Store Paradox

EVEN THOSE WHO KNOW BETTER PLAY THE SO-CALLED CHAIN STORE GAME IN AN IRRATIONAL MANNER.

The conception of predatory pricing had a long tradition in economics and can in fact be traced back to Adam Smith (1776). Also jurisprudence has been concerned with predatory practices for quite a long time as the Sherman Anti-Trust Act from 1890 shows distinctly. There is no agreement in the literature on whether predatory pricing can be maintained as a viable threat and whether the anti-trust prohibiting predatory prices are necessary. The crucial point in this debate is the credibility of threats. It is, therefore, important to analyze in a purely game theoretic context whether effective threatening of potential entrants by a monopolist can be rational. Obviously, this is an abstract problem which may very well be treated in a model disregarding, in many aspects of real markets and details of actual pricing policies. Such a model was presented by Selten (1978). The plausibility of effective deterrence asserted by Selten is not confirmed by the formal game-theoretic analysis of his scenario. This is what Selten calls the "Chain -Store Paradox". The Chain-Store paradox is a simple game in extensive from which produces an inconsistency between game theoretical reasoning and plausible human behavior. Well-informed player's must be expected to disobey game theoretical recommendations from Sultan original article.The Chain-Store paradox is a simple game in extensive from which produces an incorsistency between game theoretical reasoning and plausible human behavior. Well-informed player's must be expected to disobey game theoretical recommendations from Selten original article.

The scenario described by Selten as a 20-staged game tree composed of 20 identical two-person games played consecutively on the 20 markets. The kth stage game is described by following figure. The decision "enter", "stay out", "cooperative" and "aggressive" are abbreviated by "in", "out" "co" and "ag" respectively. Although the specific choice of numbers is consistent with customary conceptions as to the relative sizes of monopoly and duopoly profits, for each player only the ordinal ranking of the outcomes independent of the opponents outcomes are important.

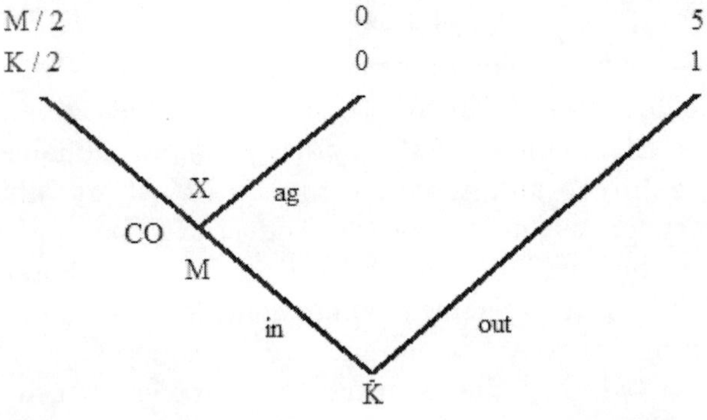

If M is to make choice in the game he knows that k chose to enter the market. As a rational profit maximize M therefore will play "co". Presuming this behavior of M it is rational for K to enter the market. Indeed, the resulting equilibrium is sequential whereas the other equilibrium, (out, ag) is not, i.e. at each node of the game tree (in, co) induces equilibrium on the case for (out, ag) at node *x*. Selten's analysis of the repeated games, the Chain-Store game, results in the assertion that the situation is the same as in the one-stage game. The unique sequential equilibrium prescribes entrance for all k ∈ (1,2,---,20) and peaceful reaction for M. The induction argument is straight Establishing the sequential equilibrium corresponds to finding an optimum in dynamic programming, the plausible "determine solution" is not supported by the analysis. This is what Selten calls the Chain - Store Paradox.

It is possible to construct a probabilistic version of the paradox. This probability theory much contain something like Harper's Popper functions constraining how an agent updates his probabilities when confronted by evidence with a zero prior probability. To generate a paradox, we need to assume the following diagram represents the structure of the game tree.

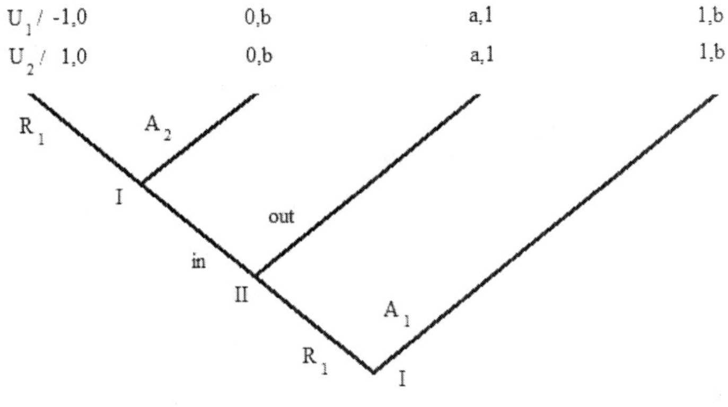

Where U_1 column represents the actual utilities of two players involved player I (the monopolist) and player II (the competitor). In fact, both players know that U_1 represents the actual utilities of both players. To generate a paradox, we assume the following six claims.

$$(a) \qquad P_c\left(\frac{U_1}{R_1}\right)=1\,iff\,P_c\left[P_m(k)\leq\frac{(a-1)}{a}\right]>0$$

$$(b) \qquad P_c\left(\frac{U_1}{R_1}\right)=0\,iff\,P_c\left[P_m(k)\leq\frac{(a-1)}{a}\right]=0$$

$$(c) \qquad P_m\left[\frac{k}{P_c\left(\frac{U_1}{R_1}\right)=1}\right]=1$$

$$(d)\ P_m\left[\frac{k}{P_c\left(\frac{U_1}{R_1}\right)=0}\right]=0$$

$$(e) \qquad p_c[(a)\,and\,b)]=1$$

$$(f) \qquad P_m\left[\frac{k}{\phi}\right]=P_c\left[\frac{k}{\phi}\right],\ for\ all\ \phi$$

The value of $\frac{(a-1)}{a}$ is the crucial value for the probability of subjunctive conditional k. If the monopolist assigns a probability of greater than $\frac{(a-1)}{a}$ to k, then he should choose A_1. Otherwise he may rationally choose R_1. It can be shown that statements a) through b) are in conflict with a fundamental with a fundamental principle of higher order probability theory, the so-called Miller's principle [Miller 1966]. Van Fraassen (1984) has produced a Dutch Book argument in favour of Miller's principle. The principle has also been endorsed by Haim Gaifman (1986) and Brian Skyrms (1980). The principle can be stated in its general form as

If $P[(\phi)\geq x \,\&\, \psi]\neq 0$, *them*

$$P\left[\frac{\phi}{P(\phi)}\geq x \,\&\, \psi\right]\geq x$$

Where ψ is any conjunction of probabilistic formulae (equations or inequalities involving only numerical constants and probability terms).

Here, we sketch briefly the Dutch book argument for Miller. Assume that $P\,[P\,(\phi)\geq x]>0$. Suppose for contradiction that $\left[\frac{\phi}{P(\phi)}\geq x\right]$ is less than x. Then the agent is vulnerable to a Dutch book. He is willing to do simultaneously the following

(a) Bet for P $(\phi)\geq$ x, at some positive odds

(b) Place a conditional bet against ϕ on the condition that $P\,(\phi)\geq$ x, at the odds x.

If $P(\phi)<x$, then the agent loses the first bet and the second conditional bet is called off. If the agent wins the first bet, then he is willing to bet for ϕ at the odds x i.e. he is willing to buy back his first bet at a net loss. We now establish that statements a) through f) are inconsistent with Miller's principle. Science we are assuming that

$$P_m\left[\frac{k}{\phi}\right]=P_c\left[\frac{k}{\phi}\right] \qquad \textit{for all } \phi$$

We replace P_m with P_c in statements c) and d). For simplicity's sake, we will simply use P. Given the competitor's certainty of statements a) and b), we can replace in c) and d) the proposition $P\left(\frac{U_1}{R_1}\right)=0$ by the proposition $P\left[P(k)\leq\frac{1}{a}\right]=0$ and we can replace the proposition $P\left(\frac{U_1}{R_1}\right)=1$ by the proposition $P\left[P(k)\leq\frac{1}{a}\right]>0.$. Let θ abbreviate $P\left[P(k)\leq\frac{a-1}{a}\right]$ Thus, statements c) and d) can restated as

c) $P\left[\frac{k}{\theta} > 0\right] = 1$

d) $P\left[\frac{k}{\theta} > 0\right] = 0$

Let's begin with statement c). Using the standard rules of probability, we can calculate that

$$P\left(\frac{k}{\theta} > 0\right) = P\left[\left(\frac{k}{P(k)}\right) < \frac{1}{a} \& \theta > 0\right].\left[P\left\{P(k) \leq \frac{1}{a} | \theta > 0\right\}\right]$$
$$+P\left[\left(\frac{k}{P(k)}\right) > \frac{1}{a} \& \theta > 0\right].\left[P\left\{P(k) > \frac{1}{a} | \theta > 0\right\}\right]$$

By Miller's principle, this sum must be less than or equal to

$$\frac{1}{a}.P\left(P(k) \leq \frac{1}{a} | \theta > 0\right) + \left[P\left(P(K) \leq \frac{1}{a} | \theta > 0\right)\right]$$

Furthermore by Miller's principle, $P\left(P(k) \leq \frac{1}{a} | \theta > 0\right)$ much be greater than zero. Hence, this sum must be less than 1, contrary to statement c).. In the case of statement d), the probability calculus entails that

$$P\left(\frac{k}{\theta} = 0\right) \geq P\left[\left(\frac{k}{P(k)}\right) > \frac{1}{a} \& \theta = 0\right].\left[P\left\{P(k) > \frac{1}{a} | \theta = 0\right\}\right]$$

By Miller's principle, $P\left[\left(\frac{k}{P(k)}\right) > \frac{1}{8} \& \theta = 0\right] > \frac{a-1}{1}$ and $P\left\{P(k) > \frac{1}{a} | \theta = 0\right\} = 1.$. Hence, their product must be greater than $\left(\frac{a-1}{a}\right)$, contrary to statement d). At least one of the two conditional probabilities

$$P\left(\frac{K}{\theta}\right) > o\, and\, p\left(\frac{K}{\theta}\right) = 0.$$

must be well defined. Yet, we have shown that neither statement c) nor statement d) can be true, contrary to what we deduced from the game tree and our theory of rational decision making.

Bibliography

1. Asher, N. and Kamp, H. (1989). Self-Reference, Attitudes and Paradox, Property, Types and Meaning, G. Chierchi. B. Partee and Turner, R. (eds.), Dordrecht: Kluwer Academic, pp 85-158.
2. Bernheim, D. (1984). Rationalizable Strategic Behavior, Econometrica, 52, pp 1007-1028.
3. Bicchieri, C. (1989). Self-Refuting theories of Strategic Interaction: A paradox of common Knowledge, Erkenntnis, 30, pp 69-85.
4. Gardenfors, P. (1978). Conditionals and Change of Belief, Acta Philosophica Fennica, 30, pp 381- 404.
5. Kaplan, D. and Montague, R. (1960). A paradox Regained, Note Dame Journal af Formal Logic, 1, pp 79-90.
6. Koons, R. (1990). Doxastic paradox without self-reference, Australasians Journal of Philosophy, 68, pp 168-177.
7. Koons, R. (1991). *Paradoxes of Belief and Strategic Rationality*, New York: Cambridge University Press.
8. Kripke, S. (1975). Outline of a theory of truth, The Journal of Philosophy, 72, pp 690-716.
9. Miller, D. (1966). A paradox of information, British Journal for the Philosophy of Science, 17, pp 59-61.
10. Rosenthal, R.W. (1981). Game of perfect information, predatory pricing and the chain store paradox, Journal of Economic Theory, 25, pp 92-100.
11. Reny, P. (1988). Rationality, Common Knowledge and Theory of Game Mimeo, Department of Economics, University of Western Ontario.
12. Selten, R. (1978). The chain Store paradox, Theory and Decisions, 9, pp 127-159.
13. Van Fraassen, B. (1984). Belief and the will, Journal of Philosophy, 81, pp 231-256.

Chapter 16

Decision Making Paradox

SELECTING THE BEST DECISION-MAKING METHOD IS A DECISION PROBLEM IN ITSELF.

The decision-making paradox relates to decision-making and it was first identified by Triantaphyllou and Mann (1989). It was further elaborated in the book by Triantaplyllou (2000) on multi-criteria decision-making. Since then it has been recognized in the related literature as a fundamental paradox in multi-criteria decision analysis (MCDA) or multi-criteria decision making (MCDM) and decision — analysis. This paradox is related to the quest for determining reliable decision-making methods. The realization for this paradox comes from the rather straightforward observation that there are numerous decision-making methods each one of which claims to be the best one. Furthermore, often these methods may yield different results when they are fed with the same decision problem and data. Finding the best decision-making method leads to the formulation of a decision problem itself for which the alternatives are the decision making methods themselves.

Triantaphyllou and Mann (1989) studied the results obtained from the four different decision-making methods. The methods examined were the weighted sum model, the weighted product model, the analytic hierarchy process and the revised analytic hierarchy process. They used the two evaluative, criteria, in an attempt to find the best method. The first criterion was to see if the method accurate in a multi-dimensional situation remained accurate in a single-dimensional case. The second criterion determined the stability of the method in yielding the same outcome when a non-optimal alternative was replaced decision problems where random numbers were used for the values of the many combinations of alternatives and criteria. The results obtained by them illustrate the paradox of deciding on single best decision-making method i.e. the difficulty that always occurs when trying to compare decision methods and choose the best decision is that a paradox is reached.

In decision theory, there are generally three steps in utilizing any decision making technique involving numerical analysis of alternatives i.e.

(a) Determining the relevant criteria and alternatives
(b) Attaching numerical measures to the relative importance of the criteria and to the impacts of the alternatives on these criteria.
(c) Processing the numerical values to determine a ranking of each alternative.

To study this paradox, we study the two commonly used methods i.e. weighted sum model and weighted product model. Despite the criticism that multi-dimensional methods have received, weighted sum model is widely used. This is the simplest MCDA or MCDM method for evaluating a number of alternatives in terms of a number of decision criteria. In general, let the given MCDA problem is defined on m alternatives and n decision criteria. Assume that all the criteria are benefit criteria, i.e. the higher the values are, the better it is. Again suppose the w_j denotes the relative weight of importance of the criterion c_j and a_{ij} is the performance value of alternative A_i when it is evaluated in terms of criterion C_j. Then, the total importance of alternative A_i, denoted as A_i^* and given by

$$A_i^* = \max \sum_j^n = 1 a_{ij} W_j \; for\, i \; = 1,2,3,......,m$$

For paragon, suppose that a decision problem of this type is defined on three alternatives A_1, A_2 and A_3 each described in terms of four criteria C_1, C_2, C_3 and C_4, then for this problem suppose the decision matrix is given by

	C_1	C_2	C_3	C_4
Alts.	0.20	0.15	0.40	0.25
A_1	25	20	15	30
A_2	10	30	20	30
A_3	30	10	30	10

The relative weight of the first criterion is equal to 0.20, the relative weight for second criterion is 0.15 and so on. Similarly the value of alternative in terms of the first criterion is equal to 25, the value of the same alternative in terms of second criterion is equal to 20 and so on. Therefore

$$A_1^* = 25 \times 0.20 + 20 \times 0.15 + 15 \times 0.40 + 30 \times 0.25 = 21.50$$

Similarly

$$A_2^* = 22.00 \text{ and } A_3^* = 22.00$$

Thus, the best alternative is A_2, because it has the maximum score. Furthermore, these numerical results imply the following ranking of these three alternatives. It is found that alternative A_2 is equal to alternative A_3 which is better than the alternative A_1.

The second popular MCDA or MCDM method is weighted product model. It is similar to the pervious method. The main difference is that instead of addition in the main mathematical operation now there is multiplication. Each decision alternative is compared with the others by multiplying a number of ratios, one for each decision criterion. Each ratio is raised to the power equivalent to the relative weight of corresponding criterion. Some of the first references to this method are due to Bridgeman (1922) and Miller and Starr(1969). To understand this problem, consider the decision problem discussed, previously. Then, if one wishes to compare the two alternatives A_k and A_l (where $m \geq k, l \geq 1$) then

$$P\left(\frac{A_k}{A_l}\right) = \prod_{j=1}^{n}\left(\frac{a_{kj}}{a_{lj}}\right)^{wj} \quad for\, k,1=1,2,3,\ldots\ldots,m$$

If the ratio $\left(\frac{A_K}{A_1}\right)$ is the greater than or equal to the value *1*, then it indicates that alternative A_k is more desirable than alternative A_l. If we are interested in determining the best alternative, then the best alternative is the one that is better than or at least equal to all other alternatives, when this method is applied on the previous data, and then the following values are derived

$$P\left(\frac{A_1}{A_2}\right) = \left(\frac{25}{10}\right)^{0.20} \times \left(\frac{20}{30}\right)^{0.15} \times \left(\frac{15}{20}\right)^{0.40} \times \left(\frac{30}{30}\right)^{0.20} = 1.007$$

Similarly

$$P\left(\frac{A_1}{A_3}\right) = 1.067 > 1\, and\, P\left(\frac{A_2}{A_3}\right) = 1.059 > 1$$

Therefore, the alternative is A_1, since it is superior to all the other alternatives. Furthermore, the following ranking of all three alternatives is

A_1 is better than A_2 is better than A_3

An alternative approach with this method is for the decision maker to use products without the previous ratio i.e. to use the following variant of main formula given earlier.

$$P(A_k) = \prod_{j-1}^{n} a_{kj}^{\,wj} \quad for\ K=1,2,3,\ldots\ldots,m$$

The term *P* (A_K) denotes the total performance value of alternative A_k when all the criteria are considered simultaneously under this model. Then when the previous data are used, exactly the same ranking is derived. Due to this reason, the paradox arises that "What decision-making method should be used to choose the best decision-making method?"

Triantaphyllou and Mann (1989) demonstrated that it is impossible to determine precisely the best decision-making method, for to do so one needs to use the best decision-making method! This problem of finding the best decision-making method always reaches a "Decision-Making Paradox" which makes any attempt in solving this problem to be of limited success.

Bibliography

1. Belton, V. and Gear, T. (1983). On a short coming of Saaty's Method of Analytic Hierarchies, Omega, Dec., pp 228-230.
2. Bernroider, E.W.N. and Stix, V. (2006). On the applicability of data envelopment analysis for multi attribute decision making in the context of information system appraisals, Data Envelopment Analysis for Multiple Attribute Decision making, Communications of the IIMA, 107, 6(2), pp 107-118.
3. Bridgman, P.W. (1922). *Dimensional Analysis,* Yale University Press, New Haven.
4. Fishhurn, P.C. (1967). Additive Utilities with Incomplete Product Set Applications to Priorities and Assignments, Operations Research.
5. Miller, D.W. and Starr, M.K. (1969). *Executive Decision and Operations Research,* Prentice-Hall, Inc. Englewood Cliffs, New Jersey.
6. Saaty, T.L. (1980). *The Analytic hierarchy Process,* Mc Graw-Hill, New York.
7. Triantaphyllou (2000). *Multi-Criteria Decision Making: A Comparative Study,* Dordrecht, The Netherland: Kluwer Academic Publishers (now Springer), pp 320, ISBN 0-7923-6607-7.
8. Triantaphllou and Mann, S.H. (1989). An examination of the effectiveness of multi- dimensional decision- making method: a decision making paradox, International Journal of Decision Support System, 5, pp 303-312.

Chapter 17

False Positive Paradox

A TEST THAT IS ACCURATE THE VAST MAJORITY OF THE TIME COULD SHOW YOU HAVE A DISEASE, BUT THE PROBABILITY THAT YOU ACTUALLY HAVE IT COULD STILL BE TINY.

The false positive paradox is a counter-intuitive statistical result whereby a positive test result is more likely to be a false positive than a true positive. This occurs when the incidence of the thing you are testing for is lower than the false positive rate of the test. This phenomenon can cause confusion when declining with medical test results for very rare conditions. In other words, the false positive paradox is a statistical result where false positive tests are more probable than positive tests, occurring when the overall population has a low incidence of a condition and incidence rate is low than the false positive rate. The probability of a positive test result is determined not only by the accuracy of the test but by the characteristics of the sampled population. When the incidence, the proportion of those who have a given condition, is lower than the test's false positive rate, even tests that have a very low chance of giving a false positive in an individual case will give more false than true positive overall. So, in a society with very few infected people, fewer proportionately than the gives false positives, there will actually be more who test positive for a disease incorrectly and don't have it than those who test positive accurate and do. Several surveys taken among physicians have shown that many do not understand the paradox. That can lead to poor treatment and prevention plans, particularly when a patient does not understand it either. Even physicians who get the mathematics may be misted if a tester is moving between populations with different real rates of incidence. Moving to a different population with different incidence can increase or decrease the effect of the paradox.

The false positive paradox arises when a fairly reliable test yields fewer correct positive results than incorrect positive because the incidence of whatever is being tested is so low in the population. For example, consider 1% of population has a disease and a test for the disease is 95% accurate. That means if one gets a positive result, there is only about a 16% chance that one actually has the disease. Specifically, the chance of a correct positive result is 1% × 95% = 0.95%, the chance for a false positive result is 99% × 5% = 4.95%, the chance of having the condition with a positive result is then $\frac{0.95}{(0.95+4.95)} \approx 16\%$.

In order to calculate it statistically, the probability of actually having a true positive result when the test is positive statisticians use Baye's theorem. Mathematically the theorem is expressed as

$$P(A|B)=\frac{P(B|A)P(A)}{P(B|A)P(A)+P(B|-A)P(-A)}$$

where the event A signifies having the condition, the event B signifies having a positive test result and hence

- $P(A|B)$ represents the chance of having the condition given a positive test result.
- $P(B|A)$ represents the chance of a true positive i.e. the positive test result being correct.
- $P(A)$ represents the incidence of the condition in the population.

- *P(-A)* represents the chance of not having the conditions i.e. 1- P *(A)*.
- $P(B|-A)$ represents the chance of a false positive i.e. getting a positive test result but not actually have the condition.

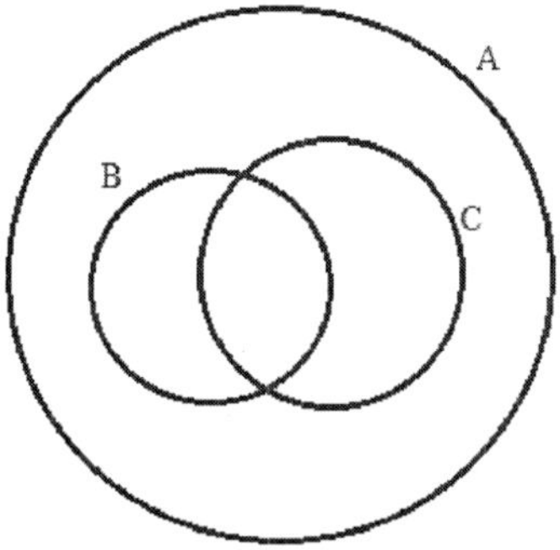

*'A'*represents the whole population to be tested, *B* represents the number of people who have the condition and *C* represents the people who get a positive test result. So it's clear that most of the people who do have the condition do get a positive result. The false positive paradox arises because *C* is at least twice as big so $B \cap C$.

In words the theorem says

$$\text{Chance that a positive result is correct} = \frac{\text{number of true positive}}{\text{number of total positive}}$$

When using the theorem it can be help us to rewrite the variables in the table format

	Has Condition *P(A)*	**Doesn't have Condition P(–A)**
Positive Test Result ***P(B)***	True Positive, *P(B/A)* (test sensitive)	False Positive, *P(B/-A)*
Negative Test Result ***P(–B)***	False Negative	True Negative (test specificity)

We now look at the real life application of this paradox, as we already said that this phenomenon can cause confusion when dealing with medical test result for very rare conditions. The result of it is that when we get a positive result, for a rare condition the actually probability that you do have the condition is still quite small ever for very accurate test. There are screening tests for many medical conditions and diseases that doctors use to diagnose people who may be at risk. For very rare diseases the vast majority of people screened will get a negative result is low. However when someone gets a positive result it does not necessarily mean that the patient actually does have the disease or condition. This causes a lot of anxiety, stress and additional financial costs for further testing. Here the false positive paradox arise because medical screening tests are never 100% specific, a screening that was specific would never give false positive and for rare disease a small percentage of a large (number of false positive) number of people can be larger than a high percentage of a small number of people (number of true positives).

Consider the example of breast cancer, the prevalence of breast cancer is approximately 0.2% in women aged over 40. The screening mammography has a sensitivity of 69% and a specificity of 94%. This gives rise to following table

	Has Breast Cancer (0.2%)	**No Breast Cancer (99.8%)**
Positive Test Result	69%	6%
Negative Test Result	31%	94%

We get

$$True\ positive = 69\% \times 0.2\% = 0.00138$$

$$False\ positive = 6\% \times 99.8\% = 0.05988$$

This is a shocking result and is the source of much controversy as to whether mammograms are even worth doing. Approximately 977 people out of 1000 people who have had a positive mammogram result do not actually have cancer. About 10% of those people will be sent for additional testing, 14% of that 10% will be sent for a biopsy, and of this group 65% will still not have breast cancer owing to the fact that the biopsy test has its own false positive rate. Another result of counter-intuitive probability states that over a decade of annual mammogram about 46% of women will receive a positive result the vast majority of which will be false positives.

Bibliography

1. Madison, B.L. (2007). Mathematical Proficiency for Citizenship, Assessing Mathematical Proficiently, Cambridge University Press, pp 122.
2. Rheinfarth, M.H. and Hewell, L.W. (1988). Probability and Statistics in Aerospace Engineering, Technical Report, p-16.
3. Smith, W. (1993). The cartoon guide to statistics, New York. Harper Collins p-49.

Chapter 18

Freedman's Paradox

DESCRIBES A PROBLEM IN MODEL SELECTION WHERE PREDICTOR VARIABLES WITH NO EXPLANATORY POWER CAN APPEAR ARTIFICIALLY IMPORTANT.

Freedman Paradox named after David Freedman, describes a problem in model selection whereby predictor variable with no explanatory power can appear artificially important. Freedman (1983) demonstrated that variable selection methods based on testing for significant F-statistics commonly include explanatory variables with no relation to the response and spuriously inflate R^2 when such irrelevant variables are present. Freedman's paradox results in spurious effects appearing important and is an extreme case of model selection bias, the bias introduced while using the data to select a single seemingly "best" model from a large set of model employing many predictor variables. Yet model selection bias also occurs when an explanatory variable has a weak relationship with the response variable. Freedman (1983) demonstrated his arguments using simulation and asymptotic calculation.

When the regression equations are used in practical world, the ratio of data points to parameters is often low, furthermore, variables with small coefficients are often dropped and the equations refitted without them. Such practice can buckle the significance level of conventional statistical tests. The existence of this effect is well known, but its magnitude may come as a surprise, even to harden statistician Freedman (1983) quantify this effect.

In simulation calculation Freedman (1983) was created a matrix with 100 rows and 51 columns. All the entries in this matrix were independent observations drawn from the standard normal distribution. In other words, this matrix was pure noise. The *51st* column was taken as the dependent variable Y in a regression equation and the first 50 columns were taken as the independent variable $X_{1,}$ $X_{2,}$......, $X_{50.}$ Then, Y was independent of X's. These data were analyzed in two successive multiple regressions and following results were obtained.

- $R^2 = 0.50$ and $P = 0.53$.
- 15 coefficients out of 50 were significant at the 25% level.
- 1coefficient out of 50 was significant at the 5% level.

Now only the 21 variables whose coefficients were significant at 25% level were allowed to enter the equation on the second pass. The results were as follows

- $R^2 = 0.36$ and $P = 5 \times 10^{-4}$.
- 14 coefficients out of 15 were significant at the 25% level.
- 6 coefficient out of 15 were significant at the 5% level.

The results from the second pass are misleading indeed for they appear to demonstrate a definite relationship between X and Y. Infect Y and selected X follow a jointly normal distribution conditioned on having significant t-statistics. The simulation was done 10 times by Freedman and found the different results.

According to Freedman (1983) an asymptotic calculation is very helpful to explain the results of the simulation experiment. He considered the model

$$Y = X\beta + \varepsilon \qquad (1)$$

where Y is an $n \times 1$ random vector, X is a constant $n \times p$ matrix with orthogonal columns, β is a $p \times 1$ vector of parameters and ε is an $n \times 1$ vector of independent normals having mean zero and common variance σ^2 i.e. $\varepsilon \sim N(0,\sigma^2)$.

Consider the null hypothesis

$$H_0: \beta = 0$$

and also $n \to \infty$ and $p \to \infty$, so that

$$\frac{P}{n} \to \rho \qquad (2)$$

where $0 < \rho < 1$.

Let us suppose that R_n^2 be the square of the conventional multiple correlation coefficient and F_n be the conventional F-statistic for testing null hypothesis. From equation (1) and equation (2), we get

$$R_n^2 \to \rho \, and \, F_n \to in \, probability$$

Now done the regression process after dropping the columns of X that fail to achieve significance at level α. Let $q_{n,a}$ be the number of remaining columns and corresponding $R_{n,a}^2$ be the new square of the conventional multiple correlation in this regression and also $F_{n,\alpha}$ be the new F-statistics.

To estimate $R_{n,a}^2$ and $F_{n,\alpha}$, which are computed by the standard formulas. For this consider z be standard normal variate and defined a function

$$\psi(z) = P \, |Z| > z$$

therefore

$$\psi(z) = \sqrt{\frac{2}{\pi}} \int_z^{\infty} \exp\left(-\frac{1}{2}u^2\right) du$$

Choose γ which is the cutoff for two-tailed Z-test at level α, so that $\psi(\gamma) = \infty$. consider

$$g(z) = \int_{|Z|>z}^{\propto} Z^2 < 1$$

Now integrate by part for $0 \leq z < \propto$, we get

$$g(z) = \psi(z) + \sqrt{\frac{2}{\pi}} \exp\left(-\frac{1}{2}u^2\right)$$

this gives

$$E\{Z^2 \| Z \ |z\} = \frac{g(z)}{\psi(z)}$$

Bibliography

1. Burnham, K. P., & Anderson, D. R. (2002). *Model Selection and Multimodel Inference: A Practical-Theoretic Approach*, 2nd ed. Springer-Verlag.
2. Freedman, D. A. (1983) "A note on screening regression equations." *The American Statistician, 37*, pp 152-155.
3. Lukacs, P. M., Burnham, K. P. & Anderson, D. R. (2010) "Model selection bias and Freedman's paradox."*Annals of the Institute of Statistical Mathematics, 62(1)*, pp 117-125.

Chapter 19

Friendship Paradox

FOR ALMOST EVERYONE, THEIR FRIENDS HAVE MORE FRIENDS THAN THEY DO.

The friendship paradox is one of a largest set of paradoxes has relevance to any variable that measures the mean size of an experience, where size refers to the number of other with whom the experience is shared. Large experience raise the mean experience size for any individuals, small experiences lower the mean experience size for only a small number of individuals. Overall, the mean size experienced by most individuals exceeds the true size. For example, most students experience a mean large then the true mean because larger classes are experienced by more students than are smaller classes.

The friendship paradox is the phenomenon first observed by the sociologist S.L. Feld (1991) that most people have fewer friends than their friends have on average. This paradox arises because extremely popular people, despite being rare, are over represented when averaging over friends. It can be explained as a form of sampling bias in which people with greater numbers of friends have increased likelihood of being observed among one's own friends. In contradiction to this, most people believe that they have more friends than their friends have. He explores mathematically the logic underlying the phenomenon, showing that the mean number of friends of friends is always greater than the mean number of friends of individuals. His further analysis shows that the proportion of individuals who have fewer friends than the mean number of friends their own friends have is affected by the exact arrangement of friendships in a social network. While it is not a mathematical necessity that each individuals will have fewer friends than the mean of his/her own friends, it is likely that most people will find themselves in this situation. Feld (1991) described the basic logic very simply. If there are some people with many friendship ties and others with few, those with many ties show up disproportionately in sets of friends. For example, these with 100 friends show up in each of 100 individual friendship networks and thus can make 100 people feel relatively deprived, while those with only one friend show up in one friendship network and can make only that one person feel relative advantaged. Hence it inevitable that individual friendship network disproportionately include those with most friends.

In general, the friendship paradox rests on the fact that the friendship counts of popular people serve as possible standards or reference point for more people that do the counts of less popular people. The following figure presents hypthetical network data to illustrate this point. Jone, who has five friends, appears in five different friendship counts, therefore, he raises the mean (or average) number of friends, friends possessed by all five of his friends. By contrast, because Mac has only one friend, Lora, he contributes to lowering the mean number of friend's friends only for Lora. As a result, Lora and Jone are the only members of this network whose own number of friends is not exceeded by the mean number of their friend's friends. In other sence, each individual is less popular than his/ her friends are Feld (1991) proves this apart from exceptional circumstances i.e. most people have fewer friends than their own friends have.

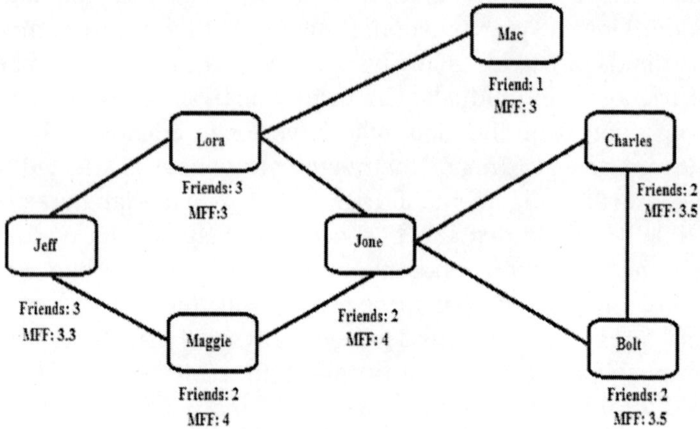

Another related aspect of the friendship paradox is that the disparity in popularity between oneself and one's friends is greater than the disparity between the self and typical or generalized others.

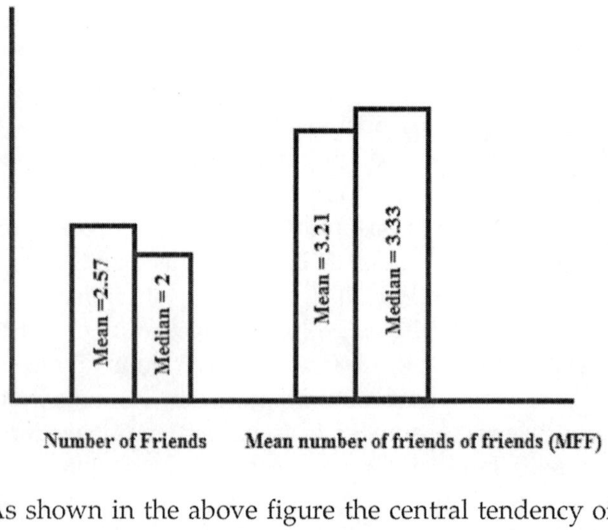

As shown in the above figure the central tendency of the number of friend's distribution of mean number of friend's friends lies to the right of that of the number of friend's distribution. This difference reflects the fact popular people increase the count of many people's "friends of friends" whereas less popular people affect fewer such counts. Due to this, one's own friends represent an objectively higher standard of comparison than does the typical population member. In spite of its apparently paradoxical nature, it can be explained as a consequence of the general mathematical properties of social networks, with the arithmetic-geometric mean inequality and Canchy-Schwarz inequality. Feld(1991) assumes that a social network is represented by an undirected graph $G = (A,B)$, where the set A of vertices corresponds to the people in the social network and the set B of edges corresponds to the friendship relation between pair of people assumes that friendship is a symmetric relation i.e. if X is a friend of Y, then Y is a friend of X. He models the average number of friends of a person in the social network as the average of degrees of the vertices in the graph. If vertex 'a' has d(a) edges touching it (a person at vertex α with number of friends), then the average number μ of friends of random person in the graph is

$$\mu = \frac{\sum_{a \in A} d(a)}{|A|} = \frac{2|B|}{|A|}$$

The average number of friends that a typically friend has can be modeled by choosing, uniformly at random and an endpoint of that edge and again calculating the degree of selected endpoint. Mathematically it is given by

$$\frac{\sum_{uv \in E} d(a)^2}{2|B|} = \mu + \frac{\sigma^2}{\mu}$$

where σ^2 is the variance of the degree in the graph, As both μ and σ^2 are positive, implies that the average degree of a friend is strictly greater than the average degree of a random node.

After this analysis, Feld (1991) goes on to make some more qualitative assumptions about the statistical correlation between the number of friends that have, based on theories of social networks such that as assortative mixing and he analyzes what these assumptions imply about the number of people whose friends have more friends than they do. Based on this analysis, he concludes that in real social networks, most people are likely to have fewer friends that the average of their friend's numbers of friends. However, this conclusion is not a mathematical certainty, there exist undirected graphs that are unlikely to arise as social network but in which most vertices have higher degree that the average of their neighbor degree.

As we said about class size, the term class size paradox can be considered a common term for all phenomena that arise when classes are of varied sizes, members of those classes disproportionately experience the large classes and most individuals therefore experience the average class size as larger then it . Such phenomena are often more than mathematical curiosities, they have implications for how people experience and respond to various aspects of their environments. As we saw that, the tendency for most people to have fewer friends than their friends have is one sociologically significant class size paradox. Individuals who find themselves associated with people with more friends than they have may conclude that they themselves are below average and somehow inadequate. The analysis presented by Feld (1991) indicated that most individuals have friends who more friends than average and so provide an unfair basis for comparison. Understanding the nature of a class size paradox or we can say friendship paradox should help people to understand that their position is relatively much better than their personal experiences have led them to believe.

Bibliography

1. Coleman, James S. (1961). *The Adolescent Society*, New York, Free Press.
2. Feld, Scott L. (1981). The Focused Organization of Society Ties, American Journal of Sociology, 86,pp 1015-1035.
3. Feld, Scott L. (1991). Why Your Friends have More Friends

Than You Do, The American Journal of Sociology, 96, pp 1464-1477.

4. Feld, Scott L. and Richared Elmore (1982). Patterns of Sociometric Choices: transitivity reconsidered. Social Psychology Quarterly, 45, pp 77-85.
5. Feld, Scott L. and Bernard Grofman (1977) Variations in Class Size, The Class Size Paradox and Consequences for Students, Research in Higher Education, 6, pp 215-222.
6. Hemenway, David (1982). Why Your Classes Are Larger than Average, Mathematics Magazine, 55, pp 162-164.
7. Hodas, N.O., kooti, F. and Lerman, K. (2013). Friends Paradox Redox: Your Friends Are More Interesting Than You, proceeding of ICWSM, 2013 at Boston, USA.

Chapter 20

Green Paradox

POLICIES INTENDING TO REDUCE FUTURE CARBON EMISSIONS MAY LEAD TO INCREASED EMISSIONS IN THE PRESENT.

For several years economists and marking analysis's have been discussing the Green paradox, a term that refers to a situation where climate policies actually worsen the problem of global warming. The rationale for the Green paradox lies in the reaction of the owners of the fossil fuel resources to climate policy measures. If resource owners expect that climate policies become stricter over time, they may decide to extract their resources faster in order to protect themselves from a bleak future. Consequently, the accumulation of atmospheric greenhouse gases occurs faster and worsens the problem of climate change. While economists in the last years have been eager in providing the theoretical contributions to Green paradox, real world climate policy making has not considered the issue. A major reason for this is that there is no empirical evidence for the appearance of green paradox. Also, the key model confirming the risk of the Green paradox relies on strong assumptions and hence the practical relevance is in doubt.

The key proponent of the Green paradox is the chief economist of the IFO-Institute in Munich, Hans- Werner Sinn.

In his contributions in newspaper and journals he warned that Kyoto countries climate policy efforts are actually leading to a Green paradox. Sinn (2008) stated that the fundamental reason for this is that most of the climate policies that are currently being enforced are aimed at reducing the demand for fossil fuel resources, such as tax on fuel consumption or a subsidy for low carbon energy supply and thereby lead to a decline in fossil fuel prices. If fossil fuel owners expect that climate policies will become stricter over time, they may decide to bring forward the sales of their fossil fuels to miss the worst of the price reductions, thereby worsening the problem of climate change.

One important caveat of the Green paradox is the assumption of an exogenously given resource stock that firms are endowed with. In reality, firms have to invest in exploration and development activities before any extraction can start. Sinn (2008) employed a basic resource extraction model based on Hotelling's theory of exhaustible resource extraction (1931) with a given fossil fuel stock, stock-depending extraction costs and no backstop technology. According to Nordhaus (1973) who coined the term, a backstop technology is a technology that constitutes a perfect substitute to the fossil fuel energy resource and is not constrained by exhaustibility. Under these assumptions, Sinn (2008) derived the effect of a carbon tax on fossil fuels and concluded that a tax that grows by more than the interest rate will generate a Green paradox. On this basis, Sinn (2008) warned that current climate policies to reduce carbon demand, ranging from carbon taxes to the development of renewable energy sources, will no mitigate climate change.

In Hotelling(1931) model, units of the resource are viewed as being available to society for extraction at any time, at a known cost that depends on the quantity extracted and possibly other factors. A striking analytic result of the model is Hotelling's rule. The rule can be proved through optimal-control analysis and is mathematically incontrovertible. The economic reasoning behind the rule is even more striking.

Consider a type-one Hotelling model of an exhaustible resource, in which extraction costs depend only on the quantity of the resource that is currently being extracted. In this case, the extraction cost of $q > 0$ units of the resource is given time by some function $\phi(q)$. This function is assumed to be increasing so that extracting more units at a given costs more and to be convex. So that it becomes ever costlier to extract additional units.

As argued by Solow (1974), the owner of a resource who wishes to maximize net present value is led to rearrange extraction such that what value is earned by the marginal unit in each time period is equal in present-value terms. If the marginal unit at one time gains less than at another time, present value can be increased by reallocating output from the period with the lower gain to the one with the higher gain. Let *p(t)* represents the price at time *t*, *r* represent the prevailing rate of interest and

$$\mu(t) = \frac{d\phi(q)}{dq} \tag{1}$$

represents the marginal cost of production. Suppose that a proposed path of extraction is such that, at times *t* and *s* during production

$$D_{(t,s)} = \frac{p(t)-\mu(t)}{(1+r)^t} - \frac{p(s)-\mu(s)}{(1+r)^s} > 0 \tag{2}$$

Then a unit of production can be re-allocated from time *s* to time *t*, at a net gain of $D_{(t,s)}$. Suppose a constant λ is determined such that for all times *t* and *s* during which extraction takes place

$$\frac{P(t)-\mu(t)}{(1+r)^t} = \lambda = \frac{p(s)-\mu(s)}{(1+r)^s} \tag{3}$$

$$\Rightarrow P(t) - \mu(t) = \lambda(1+r)^t \tag{4}$$

the net price rises at the rate of interest, Result in in equation (4) gives the Hotelling's rule.

Geneally it is assumed that the marginal cost is constant, so that for some number β

$$\phi(q) = \beta q \tag{5}$$

and hence for any value of q or t

$\mu\ (t) = \beta$

Hotelling's original paper is that he also considered what may be called a type-two model. In type two models, cost depends on the total available reserve, Q (say) as well as current output, say $\psi\ (0,Q)$. The properties of this cost function are that it is an increasing, convex function of output q and a decreasing convex function of available reserves Q. Also

$\psi\,(0, Q) = 0$ for any value of Q

A type-two model provides less sharp theoretic results than a type- one model. The rule is expressed in terms of the discounted sum, over the future of production from the resource of the increases in cost that arise because current production affects future costs through depleting total reserves. Suppose at time t during extraction, the expression for the Hotelling's rent incorporates, then

$$\sum_{s=t}^{T} \frac{\dfrac{\partial\psi\{a(s),Q(s)\}}{\partial Q}}{(1+r)^{s-t}} \tag{6}$$

Type-two model are considered to be more realistic. The function $\psi(q,\theta)$ has been a utilized empirical research in economics since the late 1970's

Even though the analysis of extraction with cost $\psi(q,Q)$ is more complicated than with cost βq. Sinn (2008) deftly uses arbitrage between adjacent for the Greeen paradox. Let

$$\psi(q,Q) = \gamma(Q)q \tag{7}$$

where $\gamma'(Q) < 0$ and $q = -\frac{dQ}{dt}$

It assumes constant marginal cost at any point in time, but increasing costs as the total stock *Q* decreases through extraction. The proceeds from a single unit of resource extracted at time t and invested for one period are $p(t)—\gamma(Q)$ of principle and $r[p(t)-\gamma(Q)]$ of interest. If instead that unit is not extracted until t+1 and also if the change in *Q* over the period is neglected as being quite small compared to *Q*, the owner gains

$$p(t+1)-\gamma(\theta)$$

Let the change in price over the period be represented by

$$\Delta p(t) \equiv p(t+1)-p(t) \tag{8}$$

The gain in two period are equal, so that

$$[p(t)-\gamma(\theta)](1+r) = p(t) + \Delta p(t)-\gamma(\theta) \tag{9}$$

Simple algebra yields a single-period generalization of Hotelling's rule

$$\frac{\Delta p(t)}{p(t)-\gamma(q)} = r \tag{10}$$

If a tax on emission is imposed and it rises through time being equal to the discounted value of the damages due to the marginal unit of emission, the result may not be what is expected by proponents of tax.

The arbitrage condition is that

$$\frac{1-\tau(t)}{(1+\delta)^t}[P(t)-\gamma(Q)](1+r) = \frac{1-\tau(0)}{(1+\delta)^{t+1}}[p(t)+\Delta p(t)-\gamma(Q)] \tag{11}$$

then the modified rule become

$$\frac{\Delta p(t)}{p(t)-\gamma(Q)} = r+\delta(1+r) \tag{12}$$

Sinn (2008) applies the following economic reasoning to this condition for different values of δ. If δ = 0, then the tax is constant and condition (10) reduces to conditions (8). Therefore, the tax is neutral, does not change the accumulation of carbon in the atmosphere. For given values of the other variables in condition (10), the change in price Δp is greater for $\delta > 0$, then for $\delta = 0$. Sinn (2008) makes technical assumptions that assure that the reserves are eventually completely exhausted in both scenarios, with and without the tax. When the reserves are exhausted the price must be the intercept of demand. Since the price rises faster with the increasing tax and since each scenario ends up with an equal total quantity extracted, the price with a tax must start out lower. In equilibrium at the current time t, therefore, output must be higher with the tax.

The economics behind the result is that the rising tax changes the relative gains to the producer over the life of the reserve, marking present extraction comparatively more attractive at the margin. Consequently, global warming is exacerbated in the present. The green paradox is that a policy designed to tax the emission of carbon in a way that is strongly related to damages caused may lead to greater current emissions. Sinn (2008) observed that other forms of policy, such as a subsidy to greener forms of energy, would be subject to a green paradox as well. His analysis, as summarized above is a benchmark for understanding the phenomenon.

Bibliography

1. Hoel, Michael (2010). Is there a Green Paradox?, CES Ifo working paper 3168.
2. Hotelling, H. (1931). The Economics of Exhaustible Resources, Journal of Political Economy, 39, pp 137-175.
3. Nordhaus, W. (1973). The Allocation of Energy Resources. Cwles Foundation paper 401.
4. Osterle, Ines (2012). Fossil Fuel Extraction and Climate Policy: A review of the Green paradox with Endogenous Resource Extraction FEEM working paper series 13.
5. Porma, Roger, Yue Ma, James McGilvray and Common (2003). Natural Resources and Environmental Economics.

3ed ed., Sddison Wesley: Pearson Education.

6. Sinn, Hans-Werneer (2008). Public Policies against global warming: a supply side approach. International Tax Public Finence, 15, pp 360-394.
7. Solow, R. M. (1974). The Economics of Resources or the Resources of Economics?, American Economic Review, 64(2), pp 1-14.

Chapter 21

Grice's Paradox

SHOWS THAT THE EXACT MEANING OF STATEMENTS INVOLVING CONDITIONALS AND PROBABILITIES IS MORE COMPLICATED THAN MAY BE OBVIOUS ON CASUAL EXAMINATION.

Grice's paradox named after Herbert Paul Grice, usually publishing under the name Paul Grice, shows that the exact meaning of statements involving conditionals and probabilities is more complicated than may be obvious on casual examination. Grice's work on the nature of meaning has influenced the philosophical study of semantics (relation between words, phrases, signs and symbols). His theory of implicature is among the most important and influential contributes to contemporary pragmatics (the way in which context contributions to meaning). Grice also distinguishes between generalized and particularized conversational implicatures. He says that particularized conversation implicatures arise in "cases in which an implicature is carried by saying that p on a periticular occasion in virture of special features about the context, cases in which there is no room the idea that implicature of this sort is normally carried by saying that *p.*"Generalized implicature arise in case in which one can say that the use of a certain form of words in an utterance would normally carry such and such an implicature or type of implicature. Grice does after full theories of generalized conversational implicatures that distinguishes them from particularized conversational implicatures and from conventional implicatures, but later philosophers and linguists, have expanded on the idea of generalized conversational implicatures. These concepts and theories are summarized in his book ***"Studies in the Way of Words"*** and he presented what he calls "Grice Paradox."

In his book, he supposes that there are two chess players (Yog and Zog) according to normal rule, but with two special conditions

- Yog as white 9 out 10 times.
- There are no draws

To date there have been 100 games and the following results are obtained

- Yog as white won 80 of 90 games
- Yog as black won zero of 10 games

Theory of probability implies that

(i) $\frac{8}{9}$ times, Yog won if Yog as white

(ii) $\frac{1}{2}$ times, Yog lost if Yog as black

(iii) $\frac{9}{10}$ that either Yog wasn't white or he won.

From these statements, it might appear one could make these deductions by contra potion and conditional disjunction

(A) If Yog as white, then $\frac{1}{2}$ of the time Yog won

(B) $\frac{9}{10}$ times, if Yog as white, then he won

But both the statements (A) and (B) are untrue, they contradict (i). In fact, (ii) and (iii) do not provide enough information to use Bayesian reasoning to each those conclusions. That might be clearer if (i) to (iii) had instead been stated like so:

(I) No information is given about when Yog as black, when Yog as white, he won $\frac{8}{9}$ times.

(II) No information is given about when Yog won, $\frac{1}{2}$ of the time, if Yog lost, Yog was black

(III) $\frac{9}{10}$ time either Yog as black and won, Yog as black and lost or Yog as white and won, no information is provided on how the $\frac{9}{10}$ is divided among those three situations.

So Grice paradox shows that the exact meaning of statements involving conditionals and probabilities are more complicated than may be obvious on casual examination.

Bibliography

1. Grice, H.P. (1969). Utterer's Meaning and Intentions, The Philosophical Review, 78, pp 86-116.
2. Grice, H.P. (1971). Intention and uncertainty, Proceeding of the British Academy, pp 263-279.
3. Grice, H.P. (1989). *Studies in the Way of Words,* Harvard University Press.

Chapter 22

Inspection Paradox

WHY ONE WILL WAIT LONGER FOR A BUS THAN ONE SHOULD.

Inspection paradox is one of the properties of renewal processes and renewal-reward processes Renewal theory is the branch of probability that generalize Poisson processes for arbitrary holding times. Mathematically, let $S_1, S_2\ S_3,\ldots\ldots$ be a square of positive independent identically distributed random variables such that

$$0 < E\,(S_i) < \propto$$

Here we refer to the random variable S_i as the ith holding time. For each $n > 0$, we define

$$J_n = \sum_{i=1}^{n} S_i$$

each J_n referred to as the nth jump time and the intervals $[J_n, J_{n+1}]$ being called renewal intervals. Than the random variable $(X_t)_{t\geq 0}$ is given by

$$X_t = \sum_{n=1}^{\alpha} \varphi\{J_n \leq t\} = \sup\{n : J_n \leq t\}$$

where φ is the indicator function and represents the number of jumps that have accorded by time t and is called a renewal process.

A curious feature of renewal process is that if we wait same predetermined time tand then observe how large the renewal interval containing t is, we should expect it to be typically large than a renewal interval of average size. Mathematically, the inspection paradox states for any $t > 0$, the renewal interval containing t is stochastically larger than the first renewal interval i.e. for all $x > 0$ and for all $t > 0$

$$P\left(S_{X_{t+1}} > x\right) \geq P\left(S_1 > x\right) = 1 - F_s\left(x\right)$$

where F_s is the cumulative distribution function of iid holding time S_j. This he calls the "The Inspection Paradox".

To understand it, suppose we know that the bus or train on a route that we take regularly does not always conform to the schedule. But while its arrival time is unreliable, it comes on average every 10 minute. But anyone who has actually experienced this has the feeling that on average the wait time is longer than that. Statistician tend to put that down to being misled by psychology, by the fact that time drags when we are waiting for something, making it seem longer than it actually is. But it turns out that psychology is not to blame, that the wait time is in fact longer than the average time between bus arrivals and this is what the inspection paradox is about. To take a concrete example, suppose that the time interval between buses on a route is 2 minutes followed by 18 minutes, then 2 minutes, ten 18 minutes and so on, so that the average time interval is 10 minutes. If one makes those on a time line, then one can easily observe that if one arrives at random, equivalent to picking a random spot on the line or inspecting it, then one is nine times more likely to arrive during the 18 minute interval than the two minute one. Hence on average, the wait time will be longer than 5 minutes. In other way the probability of arriving during 2 minutes interval is $\frac{2}{20}$ and will have an average wait time of 1 minute, while the probability of arriving during the 18 minute interval is $\frac{18}{20}$ and the average wait time will be 9 minutes. So the expected overall wait time is

$$E(W,T) = \frac{2}{20}\times 1+\frac{18}{20}\times 9 = 8.2 \; minutes$$

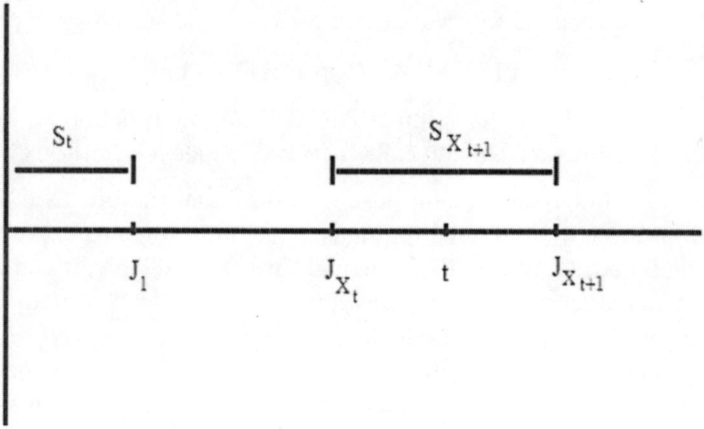

Observe that the last jump time before t is J_{x_t} and that the renewal interval containing *t* is $S_{X_{t+1}}$. Then

$$P\left(S_{xt+1} > x\right) = \int_0^\alpha \left(S_{xt+1} > x \middle| J_{xt} = s\right) f_s(s)\,ds$$

$$= \int_0^\alpha P\left(S_{X_{t+1}} > x \middle| S_{X_{t+1}} > t-s\right) f_s(s)\,ds$$

$$= \int_0^\alpha \frac{P\left(S_{X_{t+1}} > x, S_{X_{t+1}} > t-s\right)}{P\left(S_{X_{t+1}} > t-s\right)} f_s(s)\,ds$$

$$= \int_0^\alpha \frac{1-F\left[\max(x,t-s)\right]}{1-F(t-s)} f_s(s)\,ds$$

$$= \int_0^\alpha \min\left[\frac{1-F(x)}{1-F(t-s)}, \frac{1-F(t-s)}{1-F(t-s)}\right] f_s(s)\,ds$$

$$= \int_{0}^{\alpha} \min\left[\frac{1-F(x)}{1-F(t-s)}, \frac{1-F(t-s)}{1-F(t-s)}\right] f_s(s)\,ds$$

$$\geq 1-F\,(x) = \rho\,(S_1 > x)$$

Bibliography

1. Angus, J. (1997). The Inspection Paradox inequality, SIAM Review, 39, pp 95-97.
2. Cinlar, E. (1969). Markov Renewal Theory, Advance in Applied Probability, 1(2), pp 123-137.
3. Cox, D. (1970). *Renewal Theory,* London, Methuen& Co.,pp 142.
4. Doob, J.L. (1948). Renewal Theory from the Point of View of the Theory of Probability, Transactions of the American Mathematical Society, 63(3), pp 422-438.
5. Herff, W., Jochms, B. and Kamps, U. (1997). The Inspection Paradox with random time, Statistics Paper, 38, pp 103-110.
6. Ross, S.M. (2000). *Introduction to Probability Models,* New York Academic Press.
7. Ross. S.M. (2003). The Inspection Paradox, Probability in the Engineering and Informational Sciences, 17, pp 47-51.

Chapter 23

Inventor's Paradox

IT IS EASIER TO SOLVE A MORE GENERAL PROBLEM THAT COVERS THE SPECIFICS OF THE SOUGHT-AFTER SOLUTION.

The inventor's paradox is a phenomenon that occurs in seeking a solution to a given problem. Instead of solving a specific type of problem, which would seem intuitively easier, it can be easier to solve a more general problem, which covers the specifics of the sought after solution. The inventor's paradox has been used to describe phenomena in mathematics, programming and logic as well as other areas that involve critical thinking such as decision theory. In general, when you are solving a problem, you often decide to limit yourself to something that's as specific as possible. When you do so, you usually place awkward limits on developers who wish to use your framework and customers who would use it. General solutions often solve more problems than a specific solution does. There's an interesting side benefit, you can often solve the general problem with less effort, cleaner design and simpler algorithms. That's the Inventor's Paradox.

In "How to Solve It", a famous book that's quite famous in mathematics circle, George Polya introduces the Inventor's paradox "

> The most ambitious plan may have more chances of success..." provided it is not based on a more pretension but on some vision of the things beyond those immediately present"

In other words, you can frequently solve a useful general problem more effectively than a highly specialized one. It's a principle that works often in mathematics. For example, the sum of numbers sequentially from 1 to 99 is as

$$1 + 2 + 3 + 4 + \ldots\ldots\ldots\ldots\ldots\ldots + 97 + 98 + 99$$

This process, although not impossible to do in your head, can prove to be difficult for most. However, the ability to generalize the problem exists, in this case by reducing the sequence to

$$(1 + 99) + (2 + 98) + (3 + 97) + \ldots\ldots\ldots + (48 + 52) + (49 + 51) + (50)$$

Although appearing in several applications, it can be easiest to explain through inspection of a relatively simple mathematical sequence.

$$1 + 3 = 4$$

$$1 + 3 + 5 = 9$$

$$\ldots\ldots\ldots\ldots\ldots\ldots\ldots\ldots$$

$$\ldots\ldots\ldots\ldots\ldots\ldots\ldots\ldots$$

$$1 + 3 + 5 + 7 + 9 = 25$$

In allowing the sequence to expand to a point where the sum cannot be found quickly, we can simplify by finding that the sum of consecutive odd numbers follows

$$\sum_{k=1}^{n}(2k-1) = n^2$$

Once you generalize the problem in above way, you can quickly sum any sequence of number. The general problem is easier to solve.

There are many examples of the Inventor's paradox in programming. Often, the most successful frameworks are simple generalizations of complex problem. Apache web server plug-ins, Visual Basic custom controls and the Internet are but a few examples of simple generalizations of complex problem. Start with following questions:

What likely to Change?

You can't spend all of your time generalizing every block of code, but you can identify areas that you may need to future proof. It's important to generalize the way you deal with your views so you can be much better off. You are looking for ways to generalize and simplify at the same time.

Is there a different way to solve this cumbersome problem?

When we are in trouble, we usually step back and ask our self, why this way?" For example, many open source projects read configuration files as XML DOM (Domain Object Model) trees. Many developers begin to look at configuration as a Java representation of an XML file. It's not instead of looking for ways to efficiently lob DOM tress across your application, look for the reason that you are doing so may be its better to read that configuration field once, and translate it to a smaller set of concrete objects representing your configuration. You can share those at will.

The inventor's paradox represents all that's fun about programming finding simple, elegant solutions to difficult problems i.e. the general problem is easier to solve and immediately throws light on the particular one.

Bibliography

1. Bentley, Jon Louis (1982). *Writing efficient programs.* Prentice-Hall. p. 170. ISBN 0-13-970251-2.
2. Bentley, Jon Louis (2000). *Programming Pearls.* Addison-Wesley. p. 239. ISBN 0-201-10331-1.
3. Pólya, Gyorgy (1957). *How to solve it: a new aspect of mathematic method.* Doubleday. p. 253.ISBN 0-691-08097-6.
4. Tate, Bruce; Gehtland, Justin (2004). "Allow for Extension". *Better, faster, lighter Java.* O'Reilly Media, Inc. p. 243. ISBN

0-596-00676-4.

5. Welborn, Ralph; Kasten, Vincent A. (2003). "Collaborative DNA: Exploring the Dynamics". *The Jericho principle: how companies use strategic collaboration to find new sources of value.* John Wiley and Sons. p. 276. ISBN 0-471-32772-7.

Chapter 24

Kavka's Toxin Paradox

CAN ONE INTEND TO DRINK THE NON-DEADLY TOXIN, IF THE INTENTION IS THE ONLY THING NEEDED TO GET THE REWARD?

It was presented by moral and political philosopher Gregory S. Kavka in "The Toxin Puzzle" (1983) and grew out of his work in deterrence theory and mutual assured destruction. Kavka's toxin puzzle is a thought experiment about the possibility of forming and intention to perform an act which, following from reason, is an action one would not actually perform. In the January 1983 volume of the journal Analysis, Gregory Kavka presented the following philosophical thought experiment.

> "An eccentric billionaire places before you a vial of toxin that, if you drink it, will make you painfully ill for a day, but will not threaten your life or have any lasting effects. The billionaire will pay you one million dollars tomorrow morning if, at midnight tonight, you intend to drink the toxin to receive the money, in fact, the money will already be in your bank account hours before the time for drinking it arrives, if you succeed. All you have to do it is intend at midnight tonight to drink the stuff tomorrow afternoon. You are perfectly free to change your mind after receiving the money and not drink the toxin."

A billionaire will give you a million dollars if at midnight you intend to drink a toxin at noon the next day. He has a nearly infallible intention detector, so you cannot fool him. He will pay you whether or not you follow through. If you do follow through and drink the toxin you will be ill for a day, a cost that is outweighed by reward. The problem is, you know that whether or not you get the money at midnight, you will have good reason to avoid drinking the toxin at noon. Thus, assuming you are rational, you believe that you will not drink the toxin and thus you cannot intend at midnight to drink the toxin. It does not matter that it would be good to intend to drink it, you cannot coherently intend to drink it.

Basically, this puzzle or paradox represent the theories of rational decision-making due to the fact that it is incoherent to intend to do something you know you have no reason to do, or that you know you will not do and that violates expected utility theory. It's probably worth asking why this is considered an interesting problem interesting problem in the first place, since on face there might not seem to be anything difficult going on. But Kavka (1983) seems to think that it would actually be very hard, if not impossible, for someone to intend at midnight tonight to drink a painful poison tomorrow if they knew that they could eventually back out without losing any money. In asserting this he is really taking two steps

(a) to argue that a person would never actually drink the toxin, because by the time they reach the moment of truth the prize will have already been won or lost and no one would choose pain vs. not pain if all else is equal

(b) to argue that if you knew that you weren't going to drink the poison tomorrow, there is no way you could intend to tonight.

The first step is fairly sound, although one can imagine challenging it in the following way, suppose that you had somehow managed, last night, to intend to drink the poison today, isn't there then a sense in which you are committed to go through with it. If so, then it may be that to win the money you do have to drink the poison after all.

According to kavka (1983), drinking the toxin is never to your advantage regardless of whether you are paid. A rational person would know he would not drink the poison and thus could not intend to drink it

	Intend	**Do not intend**
Drink	Impossible	-10
Do not drink	Impossible	0

Gauthior (1994) argues once a person intends drinking the poison one cannot entertain ideas of not drinking

"The rational outcome of your deliberation tomorrow morning is the action that will be part of your life going as well as possible subject to the constraint that it be compatible with your commitment, in this case, compatible with the sincere intention that you form today to drink the toxin. And so the rational action is to drink the toxin."

Then according to Gauthier the pay-offs table is given by

	Intend	**Do not intend**
Drink	90	-10
Do not drink	Impossible	0

One of the central tenets of the paradox is that for a reasonable person

(a) There is reasonable grounds for that person to drink the toxin, since some reward may be obtained

(b) having come to the above conclusion there is no reasonable grounds for that person to drink to toxin, since no further reward may be obtained and no reasonable person would partake in self-harm for no benefit.

Therefore, a reasonable person must intend to drink the toxin by the first argument, yet if that person intends to drink the toxin, he is being irrational by the second argument.

The basic theory behind this paradox is that the big decision is more of the pre-decision is drinking the toxin itself. However, the pre-decision than the decision itself, we believe that the decision is drinking the toxin itself. However, the pre-decision is how we are drinking the toxin. When you think about it, the way you drink the toxin, or the attitude inside of you as you drink the toxin, really makes a huge a difference. It is not really much of a decision, because we do not decide. Our emotions are just the feelings we have inside of us. Whether it happens because of our surroundings or it happens because we are created that way, it happens and it's just there. Overall, we think that our attitude while we are drinking the toxin is not something that is between our hands so obviously it is not really a choice which is not really freedom.

Bibliography

1. Gauthier, David (1994). Assure and Threaten, Ethics, 104(5), pp 690-721.
2. Kavka, Gregory (1983). The Toxin Puzzle, Analysis, 43(1), pp 33-36.

Chapter 25

Lindley Paradox

TINY ERRORS IN THE NULL HYPOTHESIS ARE MAGNIFIED WHEN LARGE DATA SETS ARE ANALYZED, LEADING TO FALSE BUT HIGHLY STATISTICALLY SIGNIFICANT RESULTS.

Lindley Paradox describes a counterintuitive situation, in statistics, in which the Bayesian and frequentist approaches to a hypothesis testing problem give opposite results for certain choices of the prior distribution. The problem of disagreement between the two approaches was discussed in Jeffreys (1948), it become known as Lindley paradox after Dennis Lindley (1957) called the disagreement a paradox.

Lindley (1957) stated and proved the original formulation of the paradox. Suppose $X_{1,}\ X_{2,}\ldots\ldots, X_n$ compose a sequence of independent and identically distributed normal random variables with distribution $\left(x_i \middle| \theta \sim N\left(\theta,\sigma^2\right)\right)$. Here θ is an unknown location parameter of interest and σ^2 is a known constant representing the variance of the random measurements about the unknown θ. Prior option about θ is a mixture of point mass p at a specified value of θ, say θ_0 and the remaining weight 1-p distributed uniformly over an interval of width I, centered at θ_0. The well-known critical values of the sample mean $\bar{x}$, based n observations that would imply the rejection of null hypothesis $H_0 : \theta = \theta_0$ in favor of the two sided alternative hypothesis $H_1 : \theta \neq \theta_0$ at significance level α are

$$\bar{x} = \theta_0 \pm \frac{\sigma}{\sqrt{n}} Z_{\alpha/2}$$

Here $Z_{\alpha/2}$ denotes the ordinate demarking the $\frac{\alpha}{2}$ tail area for a standard normal density. The posterior probability that the null-hypothesis is true conditional upon the observation of data that achieves this significance level approaches 1 as the sample size n becomes arbitrarily large. It was proved by Lindley in 1957 as

$$P\left(H_0 \middle| \bar{X} = \bar{x} = \theta_0 \pm \frac{\sigma}{\sqrt{n}} Z_{\alpha/2}\right) = \frac{P(H_0)P(Data|H_0)}{P(H_0)P(Data|H_0) + P(H_1)P(Data|H_1)}$$

$$= \frac{p \frac{n}{\sqrt{2\pi\sigma}} e^{-\frac{1}{2}\left[\frac{\sqrt{n}}{\sigma}(\bar{x}-\theta_0)\right]^2}}{p \frac{\sqrt{n}}{\sqrt{2\pi\sigma}} e^{-\frac{1}{2}\left[\frac{\sqrt{n}}{\sigma}(\bar{x}-\theta_0)\right]^2} + (1-p)\int_{\theta_0-\frac{1}{2}}^{\theta_0+\frac{1}{2}} \frac{\sqrt{n}}{\sqrt{2\pi\sigma}} e^{-\frac{1}{2}\left[\frac{\sqrt{n}}{\sigma}(\bar{x}-\theta_0)\right]^2 \times \frac{1}{1}} d\theta}$$

$$= \frac{p e^{-\frac{1}{2}Z^2_{\alpha/2}}}{p e^{-\frac{1}{2}z^2_{\alpha/2}} + \frac{(1-p)}{1}\int_{\theta_0-\frac{1}{2}}^{\theta_0+\frac{1}{2}} e^{-\frac{1}{2}\left[\frac{\sqrt{n}}{\sigma}(\bar{x}-\theta_0)\right]^2} d\theta}$$

$$\geq \frac{p e^{-\frac{1}{2}z^2_{\alpha/2}}}{p e^{-\frac{1}{2}Z^2_{\alpha/2}} + \frac{(1-p)}{1}\frac{\sqrt{2\pi\sigma}}{\sqrt{n}}} \to 1 \qquad as\ n \to \alpha$$

The second line mealy fills in details appropriate to the components of Bayes' Theorem. In the third line, the common factor $\frac{\sqrt{n}}{\sqrt{2\pi\sigma}}$ is cancelled in all terms of numerator and denominator. In the fourth line the integral is replaced by the large value of that integral if its domain of integration were the entire real line, which would be the proportionality constant appropriate to normal density for θ centered at the conditioning value of $\overline{x}$ with variance $\frac{\sigma^2}{n}$. The limiting result follows simply since $\sqrt{n}$ appears only in the second term of the denominator of the final line.

Thus arises the paradox, that upon the observation of data that the long-standard significance testing procedures would pronounce as yielding a result that is significantly different from a prescribed value at level α would yield a posterior level of belief in the prescribed value that can be close to certainty. The paradox founders on the fact that even according to the principles of classical hypothesis testing, the choice of an appropriate significance level for a test should depend on the amount of type II error that must be incurred with it. To presume that a specific level of significance has the same meaning matter what is the sample size, that is, no matter what level of type II error is incurred, is a fundamental error that has riddled the practice of so called objective statistical method for years, even to the present day.

It can be illustrated with a numerical example, consider a certain city where 49,581 boys and 48,870 girls have been born over a certain time period. The observed proportion X of male births is thus $\frac{49,581}{48,870} \approx 0.5036$. We assume the number of male births is a binomial variable with parameter θ, we are interested in testing whether, is 0.5 or some other value i.e. null hypothesis is H_0: $\theta = 0.5$ against alternative hypothesis H_0 : $\theta \neq 0.5$. The frequentist approach to testing H_0 is to compute a p-value, the probability of observing a function of boys at least as large as X assuming H_0 is true. Because the number of births is very large, we can use a normal approximation for the fraction of male birth $X \sim N(\mu, \sigma^2)$ with

$$\mu = np = n\theta = 98451 \times 0.5 = 49225.50$$

and

$$\sigma^2 = n\theta\,(1-\theta) = 98451 \times 0.5 \times 0.5 = 24612.75$$

therefore

$$P\left(X \geq x \mid \mu = 49222.50\right) = \int_{49581}^{98451} \frac{1}{\sqrt{2\pi\sigma^2}} e^{-\frac{1}{2}\frac{(\mu-49225.5)^2}{24612.75}} du$$

$$= \int_{49581}^{98451} \frac{1}{\sqrt{2\pi(24,612.75)}} e^{-\frac{1}{2}\frac{(\mu-49225.5)^2}{24612.75}} du$$

$$\approx 0.0117$$

We would have been equally saw if we had seen 49581 female births i.e. $x \approx 0.4964$, so frequentist would usually perform a two-sided test, for which the p-value would be $p \approx 2 \times\times 0.0117 = 0.0235$. In both cases, the p-value is lower than the significance level of 5%, so the freuentist approach reject H_0 as disagreeing with the observed data.

According to the Bayesian approach, assign prior probabilities $\pi(H_0) = \pi\ (H_1) = 0.5$ and then to compute the posterior probability of H_0 using Bayes Theorem, we have

$$P(H_0 \mid K) = \frac{P(K \mid H_0)\pi(H_0)}{P(K \mid H_0)\pi(H_0) + P(K \mid H_1)\pi(H_1)}$$

After observing k = 49581 boys out of n = 98451 births, we compute the posterior probability of each hypothesis using probability mass function for a binomial variable.

$$P(K \mid H_0) = \binom{n}{k}(0.5)^k (1-0.5)^{n-k} \approx 1.95 \times 10^{-4}$$

$$P(K \mid H_1) = \int_0^1 \binom{n}{K} u^k (1-u)^{n-k} du$$

$$= \binom{n}{K} B(K+1, n-K+1)$$

$$\approx 1.02 \times 10^{-5}$$

Where B(a,b) is the Beta function. From these values, we find the posterior probability of $P(K \mid H_0) \approx 0.95$, which strongly favor H_0 over H_1. These two approaches give the different results far same problem. The Bayesian and the frequintist approaches appear to be in conflict and this is the paradox.

Bibliography

1. Deely, J. (2003). Comparing Two Group Or Treatments -A Bayesian approach in Applied Bayesian Statistical Studies in Biology and Medicine, Di Bacco, M., D' Amore,G., and Scal fari F. (eds) Boston: Kluwer Academic publishers.
2. Edward, W., Lindman, H. and Savage, L.J. (1963) Bayesian Statistical Inference for Psychological Research, Psychological Review, 70, pp 193-242.
3. Jeffreys, H. (1948). *Theory of probability*. Second Edition, Oxford: Clarendon Press.
4. Lindley, D.V. (1957) A Statistical Paradox, Biometrika, 49, pp 187-82.

Chapter 26

Low-Birth-Weight Paradox

LOW BIRTH WEIGHT AND MOTHERS WHO SMOKE CONTRIBUTE TO A HIGHER MORTALITY RATE. BABIES OF SMOKERS HAVE LOWER AVERAGE BIRTH WEIGHT, BUT LOW BIRTH WEIGHT BABIES BORN TO SMOKERS HAVE A LOWER MORTALITY RATE THAN OTHER LOW BIRTH WEIGHT BABIES. THIS IS A SPECIAL CASE OF SIMPSON'S PARADOX.

The low-birth-weight paradox is an apparently paradoxical observation relating to birth weights and mortality rate of children born to tobacco smoking mother, low birth weight children born to smoking mother have a lower infant mortality rate than the low birth weight children of non-smokers. In general, babies weighing less than a certain amount which is different for different countries have been classified as having low birth weight. In a given population, low birth weight babies have a significantly higher mortality rate than others. Therefore population with a higher rate of low birth weights typically also has higher rates of child mortality than other population. Based on prior research, the children of smoking mothers are more likely to be of low birth weight than children of non-smoking mothers. Therefore, by extension the child mortality rate should be higher among children of smoking mothers. So it is a surprising real world observation that low birth weight babies of smoking mother have lower child mortality than birth weight babies of non-smoking.

At first sight these finding seemed to suggest that, at least for some babies, having a smoking mother might be beneficial to one's health. However the paradox can be explained statistically by uncovering a lurking variable (or confounding variable) between smoking and the two key variables (birth weight and risk of mortality). Both variables are acted on independently by smoking and other adverse conditions, birth weight is lowered and risk of mortality increases. However each condition does not necessarily after both variables to the same extent. The birth weight distribution for children of smoking mother is shifted to lower weight by their mothers' actions. Therefore, healthy babies, who would weigh more if it were not for the fact their mother smoked, are born underweight. However, they still have a lower mortality rate than children who have other medical reasons why they are born underweight.

The general approach to the paradox is not limited to smoking and infant mortality, when studying the association between any prenatal variable (sex, altitude, race, multiple pregnancies etc.), adjustment for any variable (birth weight, gestational age) affected by the exposure of interest can introduce bias if the causal relation between these variables resembles that represented in following figure.

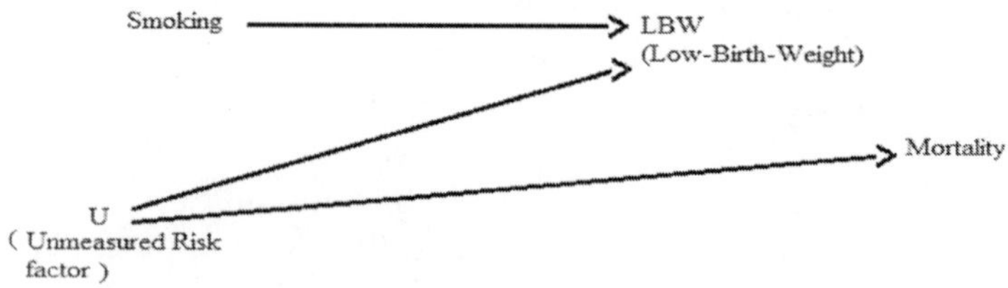

For example, in words of MacMahon et al. (1965) " at any given weight the female infant in the series with lower mean weight will have, relative to moles, a smaller proportion of member whose weight is reduced by those factors that are associated with increased mortality and the group will consistently have more favorable mortality rate." Consider the association between altitude and birth weight. Infant born at high altitudes have lower birth weights than but similar mortality rates as infants born at lower altitudes in comparable population. However, a rate ratio stratified on birth weight, unlike. The crude rate ratio, might indicate a spurious protective effect of altitude on infant mortality for LBW infant. For example, in Denver Colorado, LBW babies are born LBW because of either the altitude or other factors. However, in Los Angeles, California, all LBW babies born LBW only because of factors other than the altitude. Since the LBW caused by altitude does not seem to increase mortality, a proportion of LBW babies in Denver will not have increased mortality, while up to 100% of LBW babies in Las Angeles might have increased mortality. That would result in higher mortality for LBW babies in Los Angeles compared with LBW babies in Denver (Hernandez-Diaz et al., 2006).

If one corrected and adjusts for the confounding by smoking, via stratification or multivariable regression modeling to statistically control for smoking, one finds that the association between birth weight and mortality maybe attenuated towards the null. Most epidemiologic studies of birth weight and mortality have controlled for maternal smoking and adjusted results, although attenuated after adjusting for smoking, still indicated a significant association. In short, smoking may be harmful in that it contributes to low birth weight, but other causes of low birth weight are generally more harmful.

Bibliography

1. Hernandez-Diaz, S., Schisterman,E.F. and Herman, M.A. (2006). The Birth weight "paradox" uncovered?, American Journal of Epidemiology, 164, pp 1115-1120.
2. MacDorman, M., Atkinson, J. (1998). Infant mortality statistics form the linked birth/infant death set-1995 period data, Month Vital Statistics Report, 46, pp 1-20.

3. MacMahon, B., Alpert, M. and Salber E.J. (1965). Infant weight and parental smoking habits, American Journal of Epidemiology, 82, pp 247-261.
4. Wilcox, A. (2001). On the importance and the unimportance of birth weight, International Journal of Epidemiology, 30, pp 1233-1241.
5. Wilcox A.J. (1993). Birth weight and perinatal mortality: the effect of maternal smoking, American Journal of Epidemiology, 137, pp 1098-1104.
6. Wilcox, A.J., and Russel, I.T. (1983). Perinatal mortality: standardizing for birthweight is biased, American Journal of Epidemiology, 118, pp 857-864.

Chapter 27

Monty-Hall Problem

AN UNINTUITIVE CONSEQUENCE OF CONDITIONAL PROBABILITY.

The Monty-Hall problem is a probability puzzle. The Monty-Hall problem's origin is from the T.V show, "Let's Make A Deal" hosted by Monty-Hall. The statement of the problem is "Suppose you are in a game show and you are a contestant in a game show in which a prize is hidden behind one of three doors, you will win the prize if you select the correct door. After you have picked one door but before the door is lifted, the emcee lifts one of the other doors revealing an empty stage and asks if you would like to switch from your current selection to the remaining door. How will your changes change if you switch?": words by Mounty-Hall.

The famous question was originally proposed by a reader of "Ask Marilyn" a column by Marilyn Vos Savant in Parade Magazine in 1990 and her solution caused an uproar among Mathematicians as the answer to the problem is unintuitive, while most people would request that switching should not matter, the contestant's chances for wining in fact double if she switches curtains. Many readers of Vas Savant's column refused to believe switching is beneficial despite her explanation. After the problem appeared in Parade, approximate 10,000 readers, including near 1000 with Ph.D.

wrote to the magazine, most of them claiming Vos Savant was wrong (Tierney 1991). Even when given explanations, simulation and formal mathematical proofs, many people still do not accept that switching is the best strategy. Paul Erdos, one of the most prolific statisticians and/or mathematicians in history, remained unconvinced until he was shown a computer simulation confirming the predicted result (Vazsonyi 1999).

The Monty-Hall problem has attracted academic interest from the surprising result and simple formulation. Variations of the Monty-Hall problem are easily made by changing the implied assumptions, creating drastically different consequences. If Monty-Hall only offers the contestant a chance to switch when having initially chosen the car (behind other doors there or goats), then the contestant should never switch. The problem is a paradox of the veridical type, because the correct result is so counterintuitive if can seem absurd, but is nevertheless demonstrably true. The Monty-Hall problem is mathematically closely related to the earlier Bertrand's Box paradox.

Vas Savant's response to the problem was "Yes, you should switch. The first door has a $\frac{1}{3}$ chance of winning but the second door has a $\frac{2}{3}$ chance. Here's a good way to visualize what happened. Suppose there are a million doors and you pick #1. Then the host, who knows what's behind the doors and will always avoid the one with the prize, opens them all except door #9,99,999. You'd switch to that door pertly fast, wouldn't you? According to Carlton (2005), players initially have a $\frac{2}{3}$ chance of picking a goat. There who swap always gets the opposite of their original choice, so those who swap have $\frac{2}{3}$ chance of winning the car. The correct answer, that players who swap have a $\frac{2}{3}$ chance of winning the car. The correct answer, that players who swap have a $\frac{2}{3}$ chance of winning the car and players who stick have a $\frac{1}{3}$ chance of winning the car is based on the premise that the host knows which door hides the car and intentionally reveals a goat. If the player selected the door hiding the car then both reaming doors hide goats and the host may choose either door at random and switching door losses. In other sense, if the player initially selected a door that hides a goat, then the host's choice is no longer at random as he has no option but to show the other goat, the second goat and switching door wins for sure.

The solution presented by Vos Savant (1990) in Parade show the three possible arrangements of one car and two goats behind three doors and the result of staying or switching after initially picking door 1 in each case

Behind Door 1	Behind Door 2	Behind Door 3	Result if staying at Door #1	Result if switching to the door offered
Car	Goat	Goat	Car	Goat
Goat	Car	Goat	Goat	Car
Goat	Goat	Car	Goat	Car

A player who starts with the initial choice wins in only one out of three of these equally likely possibilities, while a player who switches wins in two out of three. The probability of winning by staying with the initial choice is $\frac{1}{3}$, while the probability of switching is $\frac{2}{3}$.

Most sources in the field of probability solve this problem by showing from first principles that the conditional probabilities [Selvin (1975), Margan et al. (1991), Chun (1991), Gilman (1992), Cortton (2005), Grinstead and Snell (2006),

Lucas et al. (2009)]. According to the definition, the conditional probability of winning by switching give the contestant initially picks door 1 and the host open door 3 is the probability the car is behind door 2 and host open door 3 divided by the probability the host open door 3. These probability table or to an equivalent decision tree

	Car Location		Host Opens	Total Probabilty	Stay	Switch
1/3	Door 1	1/2	Door 2	1/6	Car	Goat
		1/2	Door 3	1/6	Car	Goat
1/3	Door 2	1	Door 3	1/3	Goat	Car
1/3	Door 3	1	Door 2	1/3	Goat	Car

Assuming the player picks door 1, the car is behind door 2 and the host open door 3 with probability $\frac{1}{3}$. The car is behind door 1 and the host open door 3 with probability $\frac{1}{6}$. These are the only possibilities given the player picks door 1 the host open door 3. Therefore, the conditional probability of winning by switching is

$$\frac{\frac{1}{3}}{\frac{1}{3}+\frac{1}{6}} = \frac{2}{3}$$

But many probability text books and articles derive the conditional probability solution through a formal application of Bayes' theorem, like Gill (202) and Henze (1997). For this without loss of generality, let us call the curtain picked by the contestant curtain 'a', the curtain opened by Monty (host) curtain 'b' and the third curtain 'c'. We will define the following events.

- A, B and C are the events that the prize (car) is behind curtains a, b and c respectively.
- O is the event that host opens curtain b.

Then the Monty-Hall problem can be restated as

$$P(A|O) = P(C|O)$$

By Bayes' theorem

$$P(A|O) = \frac{P(A)P(O|A)}{P(O)}$$

$$P(C|O) = \frac{P(C)P(O|C)}{P(O)}$$

We assume that prize is randomly placed behind the curtains

$$P(A) = P(B) = P(C) = \frac{1}{3}$$

We compute all of the conditional probabilities for even O, as we will need them to compute *P(O)*.

- $P(O|A) = \frac{1}{2}$, if the prize is behind 'a', host can open either 'b' or 'c'.
- $P(O|B) = 0$, if the prize is behind 'b', host can only curtain 'b'.
- $P(O|C) = 1$, if the prize is behind 'c', host can only curtain 'b'.

To compute *P(O)*, we know that

$$O = (O \cap A) \cup (O \cap B) \cup (O \cap C)$$

and $(O \cap A)$, $(O \cap B)$ and $(O \cap C)$ are mutually exclusive events as the prize can be behind one curtain only, So

$$P(O) = p(O \cap A) + P(O \cap B) + P(O \cap C)$$
$$= P(A)P(O|A) + P(B)P(O|B) + P(C)P(O|C)$$

$$= \frac{1}{3}\times\frac{1}{2}+\frac{1}{3}\times 0+\frac{1}{3}\times 1=\frac{1}{2}$$

Therefore, we have

$$P(A|O)=\frac{P(A)P(O|A)}{P(O)}=\frac{1}{3}$$

$$P(C|O)=\frac{P(C)P(O|C)}{P(O)}=\frac{2}{3}$$

In other words, if we switch we double our chances of winning. We could have also arrived at this answer by noting

$$P(B|O)=0$$

any therefore $\quad P(C|O)=1-P(A|O)=\frac{2}{3}$

It is to be noted that the Monty-Hall problem is unintuitive because it is hard to see when information is brought into the problem. Mounty-Hall's opening of the curtain causes the probabilities change.

Bibliography

1. Bell, William (1992). "Comments on let's make a deal by Morgan, American Statistician, 46(3), pp 241.
2. Carlton, M. (2005). Pedigrees, Prizes Prisoners: The Misuse of Conditional Probability, Journal of Statistics Education,13(2),pp34-35
3. Chun, Y.H. (1991). "Game show Problem, OR/MS Today, 18(3), pp 9.
4. Diaconis, P. (1988). Group representation in probability and statistics, IMS Lecture Notes, Institute of Mathematical Statistics, ISBN 0-940600-14-5.
5. Eisemauer, J.G. (2001). The Mounty Hall Matrix, Teaching Statistics, 22(1), pp 17-20.
6. Gardner, M. (1959). "Mathematical Game" column, Scintific American, Oct., pp 180-182.
7. Gardner, M. (1982) *Aha! Gotcha: Paradoxes to Puzzle and Delight W.H.*, Freeman. ISBN 978-0716713616.
8. Gill, Jeft (2002). *Bayesian Methods*, pp 8-10. CRC Press, ISBN 1-58488-288-3.

9. Gillman, L. (1992). "The car and goats, American Mathematical Monthly, 99, pp 3-7.
10. Hall, Mounty (1975). The Mounty Hall Problem, letsmakeadeal.com, Includes May 12, 1975, letter to Steve Selvin.
11. Henze, Norbert (1997). *Stochastik for Einsteiger: Eine Einfuhrung in die foszinierender*, Welt des Zufalls, 9th ed 2011, pp 50-51.
12. Lucas, Stephen, Jason Rosen house and Schepler, A. (2009). The Mounty Hall problem, Reconsiderd, Mathematics Magazine, 82(5),pp75-81.
13. Morgan, J.P., Chaganty, N.R., Dahiya, R.C. and Doviak, M.J., (1991). Let's make a deal: The player's dilemma, American Statistician, 45, pp 284-287.
14. Sevin, S. (1975). A problem in probability, American Statistician, 29(1), pp 67.
15. Varzsonyi, A. (1999). Which door has the Cadillac?, Decision Line, Dec./Jan. 1999, pp 17-19.
16. Vos Savant, Marilyn (1990). "Ask Marilyn" column, Parade Magazine, pp 25.(2 Dec. 1990).
17. Vos Savant, Marilyn (1991). "Ask Marilyn" column, Parade Magazine, pp 12.(17 Feb. 1991).

Chapter 28

Morton's Fork Paradox

CHOOSING BETWEEN UNPALATABLE ALTERNATIVES.

Morton's Fork is a logical dilemma in which people are faced with two equally bad options. You, could think of it as being "between the devil and the deep blue sea", as the saying goes. Unless the victim of the dilemma manages to find an exception, the outcome of the situation will most probably be undesirable, as there can be no good outcome from any of the choices the victim is faced with. Many people find themselves struggling with Morton's Fork at some point in their lives. This term is named after Lord Chancellor John Morton, who worked in England under Henry VII in the late 15th century. According to Morton's logic, wealthy subject of the Crown obviously had money to be spared for taxes and poor subject were clearly sitting on savings, so they could also bear high taxes. Rich and poor alike found themselves at the points of "Morton's Fork" paying high taxes.

Like many logical dilemmas, Morton's Fork can have interesting implications. It sometimes comes up in game theory, with the game of bridge actually having a move called "Morton's Fork" after the dilemma which inspires it. Statisticians and people who study human behavior are often interested to see how people respond to such dilemmas and how they rationalize their behavior if the outcome turns out as poorly as expected. When faced with Morton's Fork, the temptation may be to do nothing, but sometimes this is also a bad alternative. Sometimes, a more thoughtful consideration between the options either reveals an additional choice or a choice in the array of existing options which is less repugnant. It may also be possible to subvert Morton's Fork by finding or creating an exception to the rule. Being between a rock and a hard place is sometimes solvable if one is willing to develop a hammer to smash the rock out of the way. It can also be helpful to be aware of Morton's Fork when having arguments or discussions with people over whom one has some authority. If people feel cornered by bad option, they may lash out, a result which no one wants. Parent's employers, teachers and others would be well-advised to think about the way they present options to their charges to assure them that there are choices to choose from, and that people are willing to work to make these choices as palatable as possible.

One of the most famous invocations of Morton's Fork was in Hollywood. MGM got hit by the heirs to Edgar Rice Burroughs estate for copyright infringement for doing a remake of their 1932 Tarzan movie. They could do this, because they had an agreement in 1931 to make Tarzan. The estate sued, saying that they were actually remaking the book. The presiding judge believed, and wrote in his decision, that the estate was trying to "impale" MGM with Morton's Fork, writing that "either the 1981 film followed the 1932 film, thereby infringing the book, or the 1981 film did not follow the 1932 film, thereby breaching the 1931 Agreement."

"One might perhaps have expected the plaintiffs to contend directly, in light of the issues in this lawsuit, that the 1981 film is based on the Book. However, by mounting an indirect attack, in which the major premise is that the 1932 film is based on the book plaintiffs apparently hoped to impale MGM with a "Morton's Fork" either the 1981 film followed the 1932 film, thereby infringing the Book or the 1981 film did not follow the 1932 film, thereby breaching the 1931 agreement. Even if plaintiffs' major premise were sound, which our discussion in the text, infra, demonstrates it is not, MGM was not necessarily forced into the dilemma that plaintiffs seek to create. Since the standard by which we judge the similarity of film to book in not the same standard by which we must judge the similarity between the two films, see Part III, infra, the Fork is flawed by the fact that is times are not true opposites. Thus, the possibility that the arguably infringing elements of the 1932 film in a way that did not substantially alter the story, thereby complying with the copyright law and the 1931 agreement. As it happens, this may have been the course MGM followed. Most of the specific incidents in the 1932 film that plaintiffs claim were taken from the book i.e. Holt's Killing of the ape, Tarzan's killing of the lion with a stranglehold and Holt's asking Jane if she can use a gun, are not in the 1981 film.

Bibliography

1. Watson (2013). *Civil Rights and the Paradox of Liberal Democracy*, Lexington Book, ISBN 9780739100387.
2. Whipple, T. J. and Robert (2013). *The Endurance Paradox: Bone Health for the Endurance Athlete,* Left Coast Press, pp 173-174, ISBN *9781598746174.*
3. http://www.wisegeek.com/what-is-mortons-fork.htm

Chapter 29

Navigation Paradox

Increased navigational precision may result in increased collision risk.

The Navigation paradox states that increased navigational precision may result in increased collision risk. In the case of aircraft, the advent of Global Positioning System (GPS) navigation has enabled craft to follow navigational paths with such greater precision, without better distribution of routes, co-ordination between neighboring craft and collision avoidance procedures, the likelihood of two craft occupying. The same space on the shortest distance line between two navigational points has increased. In last few decades, operational researches and statisticians have developed statistical models of processes leading to possible collisions of aircraft flying in proximity to one another in order to estimate the risk of collision. These collision risk model were applied to determine safe separation standards between pairs of co-attitude aircraft on parallel courses.

Machol (1995) discussed that the first useful collision-risk model was developed by Marks (1963) of United Kingdom's Royal Aircraft Establishment, in the context of oceaic navigation. In general, this model was based on computing the probability of proximity and conditional probability of collision given proximity purification of initial model led to a decision-making process that played a major role in arbitrating a serious disagreement over safe use of informational air-space. Reich (1966) and his co-worker modified his theory. Marks represented the aircraft parallelepiped of length 150 feet, width 150 feet and height equal to 40 feet (the tail height of aircraft). This was the first of several conservative assumptions made with the explicit assertation that all assumptions should be conservative. In this model, a collision occurred when the two boxes interested. Marks then replaced the two 150 x 150 x 40 foot boxes by a single 300 x 300 x 80 foot box repressing the other aircraft with a collision occurring when the point entered the box. Of course, statistically identical and a good bit more convenient. At that time, the quantitative aspects of establishing safe lateral separation between aircraft had been limited to considerations of the distribution of lateral deviations resulting from cross-track wander. The general assumption was that ensemble of these-errors followed a Gaussian distribution with mean zero. This primitive approach made determining a safe lateral separation value dependent on estimating the distribution's standard deviation, this estimation process was based as much on judgment as on empirical evidence. In fact, North Atlantic Ocean (NAT) navigational performance data of that time suggested that larger magnitudes of lateral error, those beyond two or three standard deviations of the empirical distribution, occurred with notably higher frequency than would be predicted from Gaussian. A handy way of referencing this higher than predicted frequency of larger errors is to refer to the ensemble of lateral errors as heavy- tailed distribution. Reich (1966) pointed out that standard deviation is a poor measure of risk, the relatively infrequent error that cause an air craft to go far into the tail of the distribution exert a markedly higher influence on risk than do these that cause the central error that contribute the overwhelming weight in determining the standard deviation. After that he recognized the "Navigational Paradox" as increases in navigational precision increase collision risk. He assumed, without empirical justification, independence of the vertical, lateral and longitudinal errors of individual aircraft and also independence of errors of proximate aircraft.

Machol (1995) talked about the another model developed by the North Atlantic System Planning Group (NATSPG) together with the error data, enabled prediction with moderate confidence of each of three risks : colliding with one's vertical, horizontal and longitudinal, neighbors. The solution eventually recommended by the SPG and adopted by International Civil Aviation Organization (ICAO) included a fourth type of neighbor. It consisted of 'composite separation' in which vertical separation remained at 2,000 feet and lateral separation at 120 miles, but additional aircraft were inserted with diagonal separation of 60 miles laterally and 1000 feet vertically. This model is used to minimize the paradoxical effect because risk of colliding with a vertical neighbor was so small that studies were made of the possibility of reducing separation from 2000 to 1000 feet. The risk of colliding with a longitudinal neighbor was also very small and longitudinal separation was later reduced to 15 minutes between same speed aircraft. The probability of collision with a lateral neighbor, which was considerably larger than other two, was determined by multiplying the probabilities of lateral vertical and longitudinal overlap. The probability of vertical overlap was surprisingly high because of the excellent vertical station-keeping. Note that if vertical station keeping is sloppy then if longitudinal and lateral separation lost, the planes will probably pass above and below each other.

Patlovany (1997) proposed a purely stochastic Monte Carlo model to compare the relative midair collision course probabilities and mean closing velocities of four systems of rules for aircraft cruising altitudes as a function of altitude error viz., current U.S. federal rules, random altitude and two proposed alternatives to the current rules. In his research paper he showed graphically that zero altitude error by pilots obeying the hemispherical cruising altitude rules resulted in six times more midair collisions than random cruising altitude non compliance. Similarly Patlovany (1997) compute model test of the Altimeter- Compass Cruising Altitude Rule (ACCAR) with zero piloting altitude error, resulted in about 60% of midair collisions counted from random altitude non compliance, or 10 times fewer collisions than the internationally accepted hemispherical cruising altitude rules. In other words, Patlovany's ACCAR alternative and Paielli's Linear cruising altitude rule would reduced cruising midair collisions between 10 and 33 times, compared to the currently recognized and internationally required, hemispherical cruising altitude rules, which institutionalize the navigation paradox on a world-wide basis. The ACCAR alternative to the hemispherical cruising altitude rule could have saved much life since the "Risk Analysis" proof that the current regulations multiply midair collision risk in direct proportion to pilot accuracy in compliance.

The Namibian collision in 1997, the Japanese near-miss in 2001, the Uberlingen collision in Germany in 2002, and the Amazon collision in 2006 are all examples where human or hardware errors doomed altitude accurate pilots killed by the navigation paradox designed into the current cruising altitude rules.

Bibliography

1. Brooker, Peter (1983). Aircraft separation assurance: Systems design, Journal of Navigation, 36(1), pp 28-36.
2. Brooker, Peter (1984). Aircraft collision risk in the North Atlantic Resign, Journal of the Operatinal Research Socity, 35(8), pp 695-703.
3. Coote, Malcom A., Schraw, Geprge W and Schwab, Robert W (1993). Oceanic requirements and benefits modeling for automatic dependent surveillance. Air Traffic Control Quarterly, 1(1), pp 31-58.
4. International Civil Aviation Organization (1988). Report of the sixth meeting of the review of the general concept of

separation panel Nov.-Dec., Montreal, Quebec, Canada.

5. International Civil Aviation Organization (1992). Report of 21st meeting of the general council of the ICAO, Sept., Montreal, Quebec, Canada.
6. Machol, Robert E., (1997). An aircraft collision model, Management Science, 21(10), pp 1089-1101.
7. Machal, Robert E., (1979). Effectiveness of the Air Traffic Control System, Journal of Operetinal Research Society, 30(2), pp 113-119.
8. Machol Robert E., (1995). Thirty Years of Modeling Midair Collisions Interfaces, 25(5), pp 151-172.
9. Marks, Brian L., (1963). Air traffic control separation standards and collision risk, RAE Tech Note No. Math 91, Royal Aircraft Establishment, Farnborough, United Kingdom.
10. Morgan, M. Granger (1993). Risk analysis and management, Scientific American, 50(3), pp 32-41.
11. Paielli, Russell A. (2000). A linear Altitude Rule for Safer and More Efficient Enroute Air Traffic, Air Traffic Control Quarterly, 8(3), pp 331-338.
12. Patlovany, Robert W. (1997). U.S. Aviation Gegulations Increase Probability of Midair Collision, Risk Analysis: An International Journal, 17(2), pp 237-248.
13. Reich, Peter G.(1966) Analysis of long-range air traffic system: sepration standards- I,II,and III, Journal of Navigation, 19(1), pp 88-96; 19(2), pp 169-176; 19(3), pp 331-338.
14. Weinberg, Alvin M.(1972). Science and trans-science, Minerva, 10 (2), pp 209-222.

Chapter 30

Necktie Paradox

A WAGER BETWEEN TWO PEOPLE SEEMS TO FAVOR THEM BOTH. IT IS VERY SIMILAR IN ESSENCE TO THE TWO-ENVELOPE PARADOX.

The necktie paradox is a puzzle or paradox within subjectivist interpretation of probability theory. In the spirit of the season we thought it would be fun to discuss a Christmas related paradox. Two men are given necktie by their respective wives as a Christmas present. Over drinks they start argument over who has the cheaper necktie. They agree to have a wager over it. They will consult their wives and find out which necktie is more expensive. The terms of the bet are that the man with the more expensive necktie has to give it to the other as the prize. Now the question arise "how fair is this wager to either play?" The first man is thinking about the bet and he suddenly feel that he has made a good wager. He gives the reasons the probability of me winning or losing is an even 50:50 chance. If I lose, then I lose the value of my necktie. If I win, however, then I win more than the value of my necktie i.e. I can bet something worth Rs. X and have a 50% chance of winning something valued at more than Rs. X a winning proposition and I should definitely make this wager. The second man, also thinking about the bet, uses the same reasoning to arrive at the same conclusion.

He feels the bet is a winning proposition for him and be is happy to take the wager. This is obviously not possible that both men can be advantaged in this zero-sum game.

The necktie paradox is, in fact, a special case of the two envelope paradox. Consequently there is a way to resolve the paradox. The issue is that each is considering the wager in terms of his own tie being the more expensive and less expensive tie at the same time. Clearly, one's own tie has to be one or the other, either the more expensive or less expensive. Therefore, the reasoning should be

- If I have the more expensive tie and I make the bet I will lose my more expensive tie.
- If I have the less expensive tie and I make the bet, I will win the more expensive tie.

There is a 50:50 chance of being in either of those situations. Hence there is no advantage to either player and there is no mystery to the bet. The game will end with each man winning the more expensive tie half of the time and hence the bet is fair.

The paradox can be resolved by giving more careful consideration to what is lost in one scenario and what is won in the other one. If we assume for simplicity that the only possible necktie price are Rs. 100 and Rs. 200 and that a man has chances of having Rs. 100 or Rs. 200 necktie, then four outcomes are possible, which are equally likely, given by

Price of 1st man (in Rs.)	**Price of 2ed man (in Rs.)**	**1st man gain/loss (in Rs.)**
100	100	0
100	200	200 Gain
200	100	200 Loss
200	200	0

We observed that first man has a 50% chance of a netral outcome, a 25% chance of gaining a necktie worth Rs. 200 and 25% chance of losing a necktie worth Rs. 200. Turning to the losing and winning scenarios, if the man loses Rs. 200, then it is true that he lost the value of his necktie and if he gains Rs. 200, then it is true that he has gained more than the value of his necktie. The win and the loss are equally likely, but what we call the value of his necktie in the losing scenario is the same amount as what we call more than the value of his necktie in the winning scenario.

Technically, if the prices of the ties could in principle be arbitrarily large, then it is possible to have beliefs about their value such that learning which was the larger would not cause any change to one's own tie about the value of one's own tie was. However, if one is 100% certain that neither tie can be worth more than Rs. 10000 then knowing which is worth more changes one's expected value of both. This paradox is rephrasing of the simplest case of the two envelopes problem.

Bibliography

1 Brown, A.C. (1995). Neckties, wallets and money for nothing, Journal of Recreational Mathematics, 27(2), pp 116-122.

2 Maurice, K. (1943). *Mathematical Recreations*, George Allen & Unwin, London.

Chapter 31

Newcomb's Paradox

HOW DO YOU PLAY A GAME AGAINST AN OMNISCIENT OPPONENT?

In mathematics and statistics, Newcomb's paradox, also referred to as Newcomb's problem. In this game, the predictor selects a human being, then offers this human being two boxes. The first box, box A, is transparent and contains a thousand dollars. The second box, box B, is opaque and contains either a million dollars or nothing. You may take only box B, or you may take boxes A and B. But there's a twist, if the predictor thinks that you will take both boxes, the predictor has left box B empty, and you will receive only thousand dollars. If the predictor thinks that you will take only box B then it has placed a million dollars in box B. Before you make your choice, the predictor has already moved on to its next game, there is no possible way for the contents of box B to change after you make your decision. If you like, imagine that box B has no back, so that your friend can look inside box B, though he can't signal you in any way. Either your friend sees that it already contains nothing imagine that you have watched the predictor play a thousand such games, against people like you, some of whom two boxes and some of whom one-boxed and on each and every occasion the predictor has predicted accurately. Do you take both boxes or only box B?

The resulting dispute over Newcomb's paradox or problem echeloned the field of decision theory into two branches, causal decision theory and evidential decision theory. The evidential theorists would take only box B in Newcomb's paradox and their stance in easy to understand. Everyone who has previously taken both boxes has received a more thousand dollars and everyone who has previously taken only box B has received a million dollars. This is a simple dilemma and anyone who comes up with an elaborate reason why it is "rational" to take both boxes is just outwitting themselves. The rational chooser is the one with a million dollars. The causal theorists analyze this paradox as: because the predictor has already made its prediction and moved on to its next game, it is impossible for your choice to affect the contents of box B in any way.

Suppose you knew for a fact that box B contained a million dollars, you would then prefer the situation where you receive both boxes ($ 1,001,000) to the situation where you receive only box B ($ 1,000,000). Suppose you knew for a fact that box B were empty, you would then prefer to receive both boxes ($ 1000) to only box B ($ 0). Given that your choice is physically incapable of affecting the content of box B, the rational choice must be take both boxes. There is dominance principle, which is that if we prefer A to B given X and also prefer A to B given X' and our choice cannot causally affect X, then we should prefer A to B. How then to explain the uncomfortable fact that evidential decision theorists end up holding all money and taking Caribbean vacations, while causal decision theorists grit their teeth and go on struggling for tenure? According to causal decision theorists the Predictor has chosen to reward people for being irrational.

Newcomb's paradox is no different from a scenario in which one can decides to arbitrarily reward people who believe that the sky is green. Suppose you could make yourself believe the sky was green, would you do in exchange for a million dollars? In essence, the predictor offers you a large monetary bribe to relinquish your rationality.

The decision theorist's dispute in not just about trading arguments with in informal, but shared, common framework. For example statisticians argue over which hypothesis best explains an experiment. The causal decision theorists and evidential decision theorists have offered different statistical frameworks for defining rational decision.

Predicted Choice	Actual Choice	Payout
A and B	A and B	($ 1000)
A and B	B only	($ 0)
B only	A and B	($ 1,001,000)
B only	B only	($ 1,000,000)

Many thought experiments similar to or based on Newcomb's problem have been discussed in last fifteen years such as Meta-Newcomb problem, Weiner's Robot problem etc. Bostrom (2001); gave a problem, labels it as Meta-Newcomb' problem. In this problem you are faced with a predictor has already made its move, placed a million dollars or nothing in box B, depending on how it predicts your choice, or else the predictor is watching to see your choice and will afterward once you have irrevocably chosen your boxes, but before you open them, place a million dollars into box B, if and only if you have not taken box A. If you know that the predictor observes your choice before filling box B, there is no way of knowing the predictor makes its move before or after your decision around half the time in both cases. Now suppose there is a Meta-predictor, who has a perfect track record of predicting the predictor's choices and also your own. The Meta- predictor informs you of the following truth-functional prediction. Either you will choose A and B and predictor will make its move after you make your choice or else you will choose only B and predictor has already made its move. An evidential decision theorist is unfazed by Nick Bostrom's Meta-Newcomb problem he takes box B and walks away, pockets bulging with a million dollars. But a causal decision theorist is faced with a puzzling dilemma. If he takes boxes A and B, then the predictor's action depends physically on his decision sot the rational action is to take only box B. But if he takes only box B, then the predictor action temporally precedes and is physically independent of his decision, so the rational action is to take boxes A and B.

The other extended problem is 'Weiner's Robot Problem' discussed by Weiner (2004). He considered that your friend falls down a mineshaft. It happens that in the world there exist robots, conscious robots, which are in most ways indistinguishable from humans. Robots are so indistinguishable from humans that most people do not know whether they are robots or humans. There are only two differences between robots and humans. First, robots are programmed to rescue people whenever possible. Second, robots have special rockets in their heels that go off only when necessary to perform a rescue. So if you are a robot, you can jump into the mineshaft to rescue your friend and your heel rockets will let you him out. But if you are not robot, you must find some other way to rescue your friend, perhaps go looking for a rope, though your friend is in a bad way, with a bleeding wound that needs a tourniquet now. Statistics collected for similar incidents show that while all robots decide to jump into mineshafts, nearly all humans decide not to jump into mineshafts. Would you decide to jump down the mineshaft?

As the solution of this paradox must point out an error in one of the two arguments Craig (1988) has suggested resolution that, in a world with perfect predictor's retro causality can occur. If a person truly knows the future and that knowledge affects his or her actions, then events in future and that knowledge affects his or her actions, then events in future will be causing effects in the past. The chooser's choice will have already caused the predictors action- some have concluded that if time machines or perfect predictor's can exist, then there can be no free will and choosers will do whatever they are fated to do. Taken together, the paradox is a restatement of the old contention that free will and determinism are incompatible, since determinism enables the existence of perfect predictors. Some philosophers argue this paradox is equivalent to the grandfather paradox. Put another way, they claim the paradox presuppose a perfect predictor; implying the chooser in not free to choose, yet simultaneously presumes a choice can be debated and decided. This suggests to some that the paradox is an artifact of these contradictory assumptions. However, Nozick's exposition specifically excludes backward causation and required only that the predictions be of high accuracy not that they are absolutely certain to be correct. Burgess (2012) has explained that, given his analysis, Newcomb's problem should be seen as being a kin to the toxin puzzle or toxin paradox. This is because both problems highlight the fact that one can have a reason to intend to do something without having a reason to actually do it.

In the same context of resolutioning this paradox Wolpert and Benford (2011) have reformulated the problem as a non-co-operative game in which player set the conditional distribution in a Bayes net. It is straight forward to prove that the two strategies for which boxes to choose make mutually inconsistent assumptions for the underlying Bayes net. Depending on which Bayes net one assumes, one can derive either strategy as optional. In this there is no paradox, only unclear language that hides that fact that one is making two inconsistent assumptions. However, that paper includes a "time reversed" version of Newcomb's problem, in which the so called predictor is made after the strategy has been chosen which the authors claim is equivalent because the probability arguments make no mention to time. In that time reversed version assumption according to which one is always completely free to choose a strategy without affecting the predictor's prediction in any way is incompatible with the original statement of the problem in which the predictor is very accurate.

Bibliography

1. Bostron, N. (2001). The Meta-Newcomb Problem, analysis, 61(4), pp 309-310.
2. Burgess Simon (2004). Newcomb's problem: An unqualified resolution, Synthese, 138(2), pp 261-287.
3. Burgess Simon (2012). Newcomb's problem and its conditional evidence: A common case of confusion, Synthese, 18(1), pp 319-339.
4. Craig, Willian Lane (1988). Tachgons, Time Travel and Divine Omniscience, The Journal of Philosophy, 85(3), pp 135-150.
5. Egan, A (2007). Some Counterexample to Causal Decision Theory Philosophical Review, 116(1), pp 93-114.
6. Garder, Martin (1974). Reflection on Newcomb's Problem: A Prediction and free-will Dilemma, Scientific American, March, pp 102-108.
7. Jeffrey, R.C. (1983). *The logic* of *Decision,* 2nd ed. Chicago, Chicago University Press.
8. Ledwig, M. (2000). Newcomb's Problem. Ph.D. Diss. University of Constance.
9. Nozick, R. (1969). Newcomb's Problem and Two Principle of Choice, In Essays in Honar of Carl. G. Hempel: A tribute on the Occasion of his Sixty-Fifth Birthday, edited by Nicholas Rescher, pp 114-146.
10. Rabin, M.O. (1980) Probabilistic Algorithm for Testing Primality, Journal of Number Theory, 12(1), pp 128-138.
11. Weiner, Malt (2004) A limiting case for causal evidential decision theory, Opiniatrety: Half to Quarter Baked Thoughts (blog), Dec. 2.
12. Wolpert, D.H. and Benford, G. (2001). What does Newcomb's paradox teach us?, Synthese, 190(9), pp 1637-1646.

Chapter 32

Ostrogorski Paradox

SHOW THAT IT CAN COME IN ELECTIONS AND REFERENDUMS TO STRONG DISTORTIONS OF THE VOTER INTENT WHEN COMPLETE WILL BE VOTED ON PARTY PROGRAMS AND NOT ON INDIVIDUAL FACTUAL ISSUES.

As in the Condorcet paradox, the collective outcome is inconsistent, despite of the individual inputs being consistent.In particular, Bezembinder and Acker (1985) proved that in every instance of the Ostrogorski paradox there is an underlying Condorcet paradox. The result is not stable with respect to the level of aggregation in the Ostrogorski paradox. In 1976, the two political scientists Hans Daudt and Douglas W. Rae presented a choice paradox before Belarusian political scientist Moissei Ostrogorski. This paradox show that it can come in elections and referendums to strong distortions of the voter intent when complete will be voted on party programs and not on individual factual issues. Consider a group of individuals where each number makes a judgment on several issues. The judgments have a binary form i.e. each issue can only be accepted or rejected. The issues are not independent, the acceptance or rejection of some propositions entails, the acceptance or rejection of other propositions. Suppose now that all the interdependence between them.

Problems arise when the adopted aggregation procedure selects a collective outcome that does not satisfy the interdependencies between the propositions. This is the case of Ostrogorski paradox (Daudt and Rae 1976, Kelly 1989). In general, Ostrogorski paradox is a voting paradox traditionally studied in social choice theory.

Now for understanding the Ostrogorski paradox consider an election involving five voter (A, B, C, D and E), 2 parties (X, Y) and 3 issues (1, 2 and 3) where each voter views the issues to be of equal importance and no other consideration influence the voters opinions about the parties. Consider two ways of determining the dominant party

a) Each voter prefers the party that is closer to his opinion on more issues.
b) For each issue, the winner is the party that receives more votes than its competitor, the dominant party is the one winning on more issues than the other.

Then the Ostrogoski paradox occurs when the outcome differs in these two cases

Issue	**Issue 1**	**Issue 2**	**Issue 3**	**The voter Votes for**
Voter A	X	X	Y	X
Voter B	X	Y	X	X
Voter C	Y	X	X	X
Voter D	Y	Y	Y	Y
Voter E	Y	Y	Y	Y
Winner	Y	Y	Y	?

The above table contains the distribution of opinions about parties X and Y provides a strong version of the paradox because not only do the results differ under procedures (a) and (b), but the winner under (b) is unanimous. Replacing any one Y with an X creates a weaker version of the paradox where just a majority differs under (a) and (b).

One might worry whether the qualitative nature of the above table is misleading, precise manner. To address this concern, Nurmi and Saari (2010) represent each issue with a coordinate direction in a 3 – dimensional Euclidean space R^3, where

$$X = (4,4,4)$$

and $$Y = (3,3,3)$$

Represent each voter's stance for the three issues with an ideal point in R^3. Let us consider

A's views be at (4,5,3.2)

B's views be at (5,3.2,4)

C's views be at (3.2,4,5)

D's views be at (2,2,2)

E's views be at (1,1,1)

By comparing these component values with those of X and Y, the above table conclusion of a majority support for Y for each issue follows. In contrast, each of A, B and C is closer to X than to Y, the common distance from A, B or C to X given by the sum of distance of each issue is $0.8 + 0 + 1 = 1.8$, which is smaller than the common distance of any of these three points to Y which is 3.2. The conclusion holds with other distances, i.e. the common Euclidean distance to X is $\sqrt{1.64}$ while that to Y is the $\sqrt{5.04}$.

Nurini and Saari (2010) talked about the appropriate way to represent a voter's preferences. In above table replace voter with criterion and imagine and individual trying to determine whether to support candidate X or Y. The criteria may be relevant educational background, political experience, negotiation skills, political connections etc. The issues might be education, economy and foreign policy. Each row and column table entry indicates our voter's preference of a candidate for that criterion and issue. While there are at least two natural ways for our voter to make his choice, if all issues and criteria are deemed of equal importance, our voter's decision is ambiguous. Geometrically, our voter's views can be represented with the above values for the locations of the criteria over issues, the distance of his opinions from Y is 18.6 while that from X is 20.4. Using the Euclidean metric, the distance from Y is $\sqrt{30.04}$ while the distance from X is the larger $\sqrt{43.92}$. With either metric, then, this individual's views are closer to that of candidate Y than of candidate X. While the standard spatial voting assumption has this individual selecting the closest candidate Y, is this the correct choice? In reality, it cannot be inferred in a pair wise comparison between X and Y whether our voter would always accept Y. In fact, by resorting the reasonable principle of basing his choice on the criterion-wise performance of candidates, they will vote for X because X outperforms Y on three criteria, while Y beats X on only two.

This assertion correction corresponds to common usage where a voter justifies his support for a candidate in terms of his education, experience, skills and so forth. Because Y is closer to the individual's beliefs on each dimension, these problems cannot be resolved by assigning salience weights to issue dimensions. Static considerations, which may underlie occasional votes against preference, do not enter into the calculus dictating the choice of X rather than Y because there are only alternatives and the other voter's ideal points is not known. As a result, Ostrogorski paradox is the feature that the group members have to express a judgment or cast a vote on several interconnected items. The rules that make the items interdependent are of a different kind. The rules that make the items interdependent are of a different kind.

In this paradox the individual and collective choices on the issues determine which party the person and the group support. The value assigned to some proposition (issues) dictates the individual and collective values on other propositions. Through this paradox cannot be strictly defined as a judgment aggregation paradox, the recognition that the cause of the puzzling results resides in which order the majority rule and the specific aggregation function is applied, should make them be recognized should not be restricted to a particular structure of judgment aggregation problems, where the dependence relations are of a logical type, but be broadened to include other types of dependencies. This paradox occur because the aggregations on the premises (issues) and on the conclusion (party) go in two divergent directions. On the one hand, the premises that receive the highest degree, of support cannot consistently provide reasons for the most popular conclusion. On the other hand, the socially selected conclusion cannot be inferred from the premises that received the majority of the votes.

Bibliography

1. Bezembinder, T. and Acker P van (1985). The Ostrogorski Paradox and its relation to no transitive choice, Journal of Mathematical Sociology, 11, pp 131-158.
2. Brams, S. J., Kilgour, D.M. and Zwicker, W.S. (1998). The paradox of multiple elections, Social Choice and Welfare, 15, pp 211-236.
3. Daudt H. and Rae D.W. (1976). The Ostrogorski paradox: A peculiarity of compound majority decision, European Journal of Political Research, 4(4), pp 391-399.
4. Deb, R. and Kelsey, D. (1987). On constructing a generalized Ostrogorski paradox: necessary and sufficient condition, Mathematical Social Science, 14, pp 161-174.
5. Gehlein, W.V. (2006). *Condorcet's Paradox*, Springer Verlag, Berlin.
6. Kelly, J.S. (1989). The Ostrogorski paradox, Social choice and welfare, 6, pp 71-76.
7. Laffond, G. and Laine, J. (2006). Single-switch preferences and the Ostrogorski paradox, Mathematical Social Science, 52, pp 49-66.
8. List, C. (2005). The probability of inconsistencies in comple collective decisions, Social Choice and Welfare, 24, pp 3-32.
9. Hurmi, H. (1999). *Voting Paradoxes and How to Deal with Them*, Springer Verlag, Berlin.

Chapter 33

Parrondo's Paradox

It is possible to play two losing games alternately to eventually win.

In the statistical game theory, Parrondo's paradox has been described as a combination of losing strategies becomes a winning strategy. It is named after its creator, Juan Par rondo, who discovered the paradox in 1996. In other words this paradox is described as, there exist pairs of games, each with a higher probability of losing than winning, for which it is possible to construct a winning strategy by playing the games alternatively Parrondo devised the paradox. In general, it is described using two games. One game is a simple coin toss, a zero-order game. The other game has the structure of Markov Chain; it uses several coins, one of which is chosen for play depending on what has happened on previous coin tosses. The expected return from the zero-order game depends only on the probability of landing heads. In contrast, the expected return from the history-dependent game depends not only on the probabilities with which the various coins land heads but also on how often each coin is tossed, which is determined by steady-state probabilities of the Markov Chain.

The games are set up so that if played separately, each game has a negative expected return. Consider randomly switching between the games. Hence the answer is that randomization changes the transition probabilities and therefore, the steady state probabilities for the Markov-chain game. An unconditional probability, the probability of winning the memory less game, changing the steady-state probabilities of the latter and reversing a player's fortune. For another interpretation of the paradox consider playing only the history-dependent game, it leads to losses. Now add on option to each state of history-dependent game, allow the player, the option of tossing another coin. Surely the option is worth using only if the new coin gives better odds in at least some states of the history-dependent game. Now suppose we impose the constraint that a player must use the new coin with the same probability in all states of the history-dependent game. Then it is possible that, even if new coin offers unfavorable odds, the modified game has winning odds. The demonstration of this feasibility is Parrondo's Paradox.

The original Parrondo's game can be formed using elementary probability rule i.e. we win with a probability p or lose with probability 1-p. Such games or processes are well known and can be likened to going on a biased random walk or tossing a biased coin. Since each step of the games consists of a win-loss decision they can be formed by using a set of biased coins.

Game A is straight forward and consists of a single branch element. The probability of winning and losing is p and $1-p$ respectively Game B is a little more complicated, as a game rule is first required to chose which of two coins to toss. It is described by the following statement. If the present capital is a multiple of an integer M, then the chance of winning in p_1, if the capital is not a multiple of M, then the chance of winning is p_2. The losing probabilities are $1-p_1$ and $1-p_2$ respectively. Thus game B uses two coins, the rule of which one to use depends on the value of the current capital. For future reference let the coins B1 and B2 have winning probabilities p_1 and p_2 respectively.

Then two games are represented diagrammatically as:

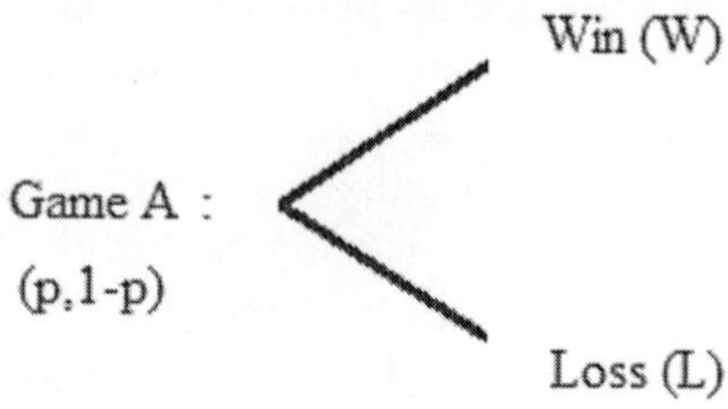

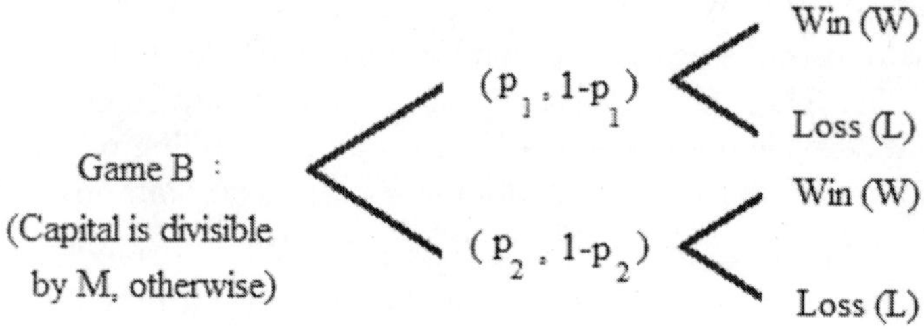

Using branching elements to represent decision conditions or win/ loss probabilities. The notation (*,*) at the left of the branch gives the probabilities or conditions for taking the top and bottom branch respectively. If we require control of the three probabilities p, p_1 and p_2 via a single variable a biasing parameter ε can be used to represent a subset of the parameter space with the transformation

$$p = \frac{1}{2} - \varepsilon$$

$$p_1 = \frac{1}{10} - \varepsilon$$

$$p_2 = \frac{3}{4} - \varepsilon$$

For a simpler example of how and why the paradox works, again consider two games Game A and Game B, this time with the rule

b) In Game A, you count simply lose Rs. 1 every time you play

b) In game B, you count how much money you have left. If it is an even number, you win Rs.3, otherwise you lose Rs.5.

Say you begin with Rs.100 in your pocket. It you start playing Game A exclusively, you will obviously lose all your money in 100 rounds. Similarly, if you decide to play Game B exclusively, you will also lose all your money in 100 rounds. However, consider playing the games alternatively, starting with Game B, followed by A, then by B and so on (BABA---). It should be easy to see that you will steadily earn a total of rupee 2 for every two games. Thus, even though each game is a losing proposition if played alone, because the results of Game B are affected by Game A, the sequence in which the games are played can affect how often Game B earns you money and subsequently the result is different from the case where either game is played by itself.

Bibliography

1. Harmer, G.P. and Abbott, D. (1999a). Losing can win by Parrondo's paradox, Nature, 402, pp 864-871.
2. Harmer, G.P. and Abbott, D. (1999b). Parrondo's Paradox, Statistical Science, 14, pp 206-213.
3. Harmer, G.P., Abbott, D. and Taylor P.G. (2000). The Paradox of Parrondo's game. Proc. Royal Socity of Landon, 456, pp 1-13.
4. Harmer, G.P., Abbott, D., Taylor P.G., Pearce C.E.M. and Parrondo, J.M.R. (2000). Information entropy and Parrondo's Discrete-time ratchet. Proc. Stochastic and Chaotic Dynamics in the lakes, Ambleside, U.K., American Institute of Physics.
5. Lyengar, R. and kohli, R. (2004) Why Parrondo's paradox is irrelevant for utility theory, stock buying and the emergence of life, Complexity, 9(1), pp 23-27.
6. Minor, D. (2003). Parrondo's Paradox-Hope for losers! The College Mathematical Journal, 34(1), pp 15-20.
7 .Stutzer, M. (2010a). The Paradox of Diversification, The Journal of Investing, 19(1), pp 32-35.

Chapter 34

Prevention Paradox

FOR ONE PERSON TO BENEFIT, MANY PEOPLE HAVE TO CHANGE THEIR BEHAVIOR, EVEN THOUGH THEY RECEIVE NO BENEFIT, OR EVEN SUFFER, FROM THE CHANGE.

The prevention paradox was first formally described in 1981 by the epidemiologist Geoffrey Rose. In general, this paradox is related to take decision which describes the seemingly contradictory situation where the majority of cases of a disease come from a population at low or moderate risk of that disease and only a minority of cases come from the high risk population (of the same disease). Rose (1981) described the distribution of risk levels for major determinants of disease follows a continuum in which the high risk person are at the extreme end. A large number of person with moderately increased risk levels. Thus, interventions targeting the general population, aiming at shifting the risk curve to the left, are more effective than interventions targeting high-risk groups. This is termed as the prevention paradox, since it is not the individuals with moderately elevated risk that have the greatest benefit from such interventions. In other words, this situation is paradoxical because it is common and logical to equate high risk populations with making up the majority of the burden of disease.

Kreitman (1986) reported evidence that the prevention paradox applies to alcohol, noting that most individuals with alcohol-related problems consumed less than the thresholds commonly used to define Hazardous drinking on average. In an Australian population-based survey, Stockwell et al. (1996) also found support for the prevention paradox, with only 25% of the 67 individuals who reported intoxication related harm exceeding the threshold for hazardous drinking on average. But the phrase "on average" here is key. Stockwell et al. (1996) also reported evidence that challenged the prevention paradox. When they defined high-risk using per occasion threshold, 87% of those 67 individuals reported hazardous drinking on the day the event occurred. Assessment of per-occasion alcohol consumption has been limited. Kreitman (1986) and Gmel et al. (2001) looked at typical frequency of heavy drinking, not at drinking on the day the event occurred.

Spurling et al. (2005) conducted a population-based case-control and case-crossover study in all 3 emergency departments in Boone country, Mo. Data were collected from 2,517 patients with and acute injury and 1856 age and sex matched controls selected by random digit dialing. Spurling et al. (2005) found the result that the population attributable fraction (PAF) associated with drinking in the 6 hours before injury, the proportion of injuries that would not have occurred in the absence of drinking, was 10.6% in case-crossover analysis and 8.5% in case control analysis. The PAF that was due to what is usually considered nonhazardous alcohol consumption was 4.5% in case-crossover analysis and 3.1% in case-control analysis. The PAF that was due to alcohol dependence was 4.0%. On the bases of these results they conclude that injury is associated more with an occasion of alcohol consumption than with alcohol dependence. Then the paradox is that substantial proportion of the PAF that is due to an occasion of alcohol consumption is from what are usually considered low-risk quantities.

In similar sense, Romelsjo and Danielsson (2012) examined the extent to which the prevention paradox applies to the relationship between alcohol consumption, heavy episodic drinking (HED) and alcohol-related problems in adolescents. They conduct a study which is based on 7288 alcohol-consuming adolescents aged 13-17 years. The proportions (in %) of problems related to drinking measures, were calculated. They considered the upper 10% and bottom 90% of drinkers by annual alcohol intake and those with frequent (monthly), less frequent and no heavy drinking episodes. They found the result that the bottom 90% of consumers by annual intake accounted for a large majority of the alcohol-related problems among boys and girls at all ages. The share of problems accounted for by monthly HED's increased with age, from ~10% among those aged 13 years to > 50% among those aged 17 years. Attributable proportions for the top 10% alcohol consumers ranged between 22% and 37%. Their analyses suggest that the prevention paradox is valid for adolescent boys and girls aged 15 years and applies to a large range of alcohol-related problems of varying severity. Their results imply that not only that prevention directed at all adolescents is essential, but also that HED should be particularly noticed.

Bibliography

1. Bruzzi, P., Green, S.B., Brinton, L.A. and Schairer, C. (1985) Estimating the population attributable risk for multiple risk factor using case- control data. American Journal of Epidemiol, 122, pp 904-914.
2. Gmel, G., klingemann, S., Muller, R. and Brenner, D. (2001). Revising the preventive paradox: The swiss case, Addiction, 96, pp 273-284.
3. Heller, R.F., Buchan, I., Edwards, R., Lyratzopoulas, G., McElduff P. and St Leger, S. (2003). Communicating risks at the population level: application of population impact numbers, British Medical Journal, 327, pp 1162-1165.
4. Kreitman, N. (1986) Alcohol consumption and the preventive paradox, British Journal of Addiction, 81, pp 353-363.
5. Romelsjo, A. and Danielsson, A. K. (2012). Does the

prevention paradox apply to various alcohols and problem among Swedish adolessents, European Journal of Public Health, 22(6), pp 899-90.

6. Rose G. (1981) Strategy of prevention: lessons from cardiovascular disease. British Medical Journal, 282, pp 1847-1851.
7. Rose, G. (1985). Sick individuals and sick population. Intervational, Journal of Epidemiology, 14, pp 32-38.
8. Sinclair, J.D., Sillanaukee, P. (1993). The preventive paradox: A critical examination. Addiction, 88, pp 591-595.
9. Sparling, M.C. and Vinson D.C. (2005). Alcohol-Related Jujuries: Evidence for the prevention paradox, Annals of Family Medicine, 3(1), pp 47-52.

Chapter 35

Proebsting's Paradox

THE KELLY CRITERION IS AN OFTEN OPTIMAL STRATEGY FOR MAXIMIZING PROFIT IN THE LONG RUN. PROEBSTING'S PARADOX APPARENTLY SHOWS THAT THE KELLY CRITERION CAN LEAD TO RUIN.

Proebsting paradox is an argument that appears to show that the betting rule known as the Kelly criterion can lead a bettor to risk an arbitrarily high proportion of his wealth on the outcome of a single event. According to this criterion one ought to choose the size of one's bet so as to maximize the expected growth of one's wealth. Addressing the paradox is important in that it seems to contradict the well-known fact that a bettor that follows the Kelly criterion can never be ruined absolutely or asymptotically. The paradox was first communicated by **Todd Proebsting** to Ed Thorp by email, which in turn made it publicly known in an article in the September 2008 issue of Wilmott Magazine, a magazine that server the quantitative finance community.

Suppose that you believe an event will occur with 50% probability and somebody offers you 2 to 1 odds on that event. How much money you should place on this bet depends on how you feel about the tradeoff between risks and reward that is being offered to you. If you were neutral to risk and all you cared about was the expected final value of wealth, then you would place 100% of your wealth on the gamble and you would ignore the fact that you may end up losing everything with probability 50%. The Kelly criterion would have you be much more conservative than that in that it would instead have you focus on the expected growth rate of your wealth. According to Poundstone (2005) the Kelly strategy's promise of doing better than any other strategy in the long run seems compelling, some economists have argued strenuously against it, mainly because an individual's specific investing constraints may override the desire for optimal growth rate. The conventional alternative is expected theory which says bets should be sized to maximize the expected utility of the outcome. Thorp (2008a) talked about the Kelly supporters usually argue for fractional Kelly, betting a fixed fraction of the amount recommended by Kelly, for a variety of practical reasons, such as wishing to reduce volatility, or protecting against non-deterministic errors in their advantage calculations. In general, for simple bets with two outcomes, one involving losing the entire amount bet, and the other involving winning the bet amount multiplied by the payoff odds, then the Kelly bet is

$$f^* = \frac{bp - q}{b} = \frac{p(b+1) - 1}{b}$$

Where f^* is the fraction of the current wealth, b is the net odds received on the b to 1, p is the probability of winning and q is the probability of losing,which is 1-p.

In the case in which you are offered the 2 to 1 odds, say situation A, the task is to find fraction f_W of your wealth that solves the problem

$$\max_{f} \; 0.5 \ln (W - f_W) + 0.5 \ln (W + 2 f_W)$$

where f_W is the function wealth W, which yields $f_W = 0.25$ and exposes you to a for lower risk, then if you place all your money on the bet. More generally, if you are offered 50 : 50 bet that pays b to 1 the Kelly criterion would have you bet a fraction of your wealth equal to

$$f^* = \frac{b-1}{2b}$$

Hence, if you were offered 5 to 1 odds, say Situation *B*, according to the Kelly criterion you would place $f_B = 40\%$ on your wealth on the bet.

Now assume that these bets occur in sequence. You are offered 2 to 1 odds, bet 25% of your wealth and then or offered 5 to 1 odds, Situation *C*. The Kelly criterion will indeed have you place an additional bet, which can be computed as

$$\max_f \; 0.5 \ln (W-0.25\,W-f_W) + 0.5 \ln (W+2 \times 0.25W + 5W)$$

which yield $f_* = 0.225$. The paradox is that total bet in this situation

$$f_C = 0.25 + 0.225 = 0.475$$

is larger than the 0.4 Kelly fration if the 5 to 1 odds are offered from the beginning. It is counterintuitive that you bet more when some of the bet is at unfavorable odds. Moreover, Thorp (2008b) showed that if a gambler is offered 2 to 1 odds then 4 to 1, then 8 to 1 and so on , the Kelly criterion would have you eventually bet your entire wealth, therefore exposing the bettor to risk of complete ruin of exactly 50%, just as if he/she was risk neutral.

The easy way to conclude the paradox is to note that Kelly assumes that odds do not change. A Kelly bettor who knows odds might change should factor this into more complex Kelly bet. For example, lets us suppose a Kelly bettor is given a onetime opportunity to bet a 50:50 proportion at odds of 2to 1. He knows there is 50% chance that a second onetime opportunity will be offered at 5 to 1 odds. Then two function f_1 and f_2, we should

$$\max_f \; 0.25 \ln (1+2f_1) + 0.25 \ln (1-f_1) + 0.25 \ln (1+2f_1 + 5f_2) + 0.25 \ln (1-f_1-f_2)$$

The answer turns out to be zero at 2 to 1 odds and wait

for the chance of betting at 5 to 1, in which case you bet 40% of wealth. If the probability of being offered is 5 to 1 odds less than 50%, then some amount between zero and 25% will be bet at 2 to 1. If the probability of being offered is 5 to 1 odds more than 50%, then the Kelly bettor will actually make a negative bet 2 to 1 odds. Predominantly, the Kelly criterion is supposed to do bettor than any essentially different strategy in the long run and have zero change of ruin, as long as the bettor knows the probabilities and payouts. In fact, it can be frustrated by unexpected new offers is puzzling. It is also puzzling that the Kelly bettor bets more at blended 2 to 1 and 5 to 1 odds than at 5 to 1 odds and that is improving odds that lead to the possibility of ruin.

Based on an argument along the line of what was espoused about, Aaron Brown argued that this analysis make it clear that the change in behavior results from the mark-to-market loss the investor experiences when the new payout is offered. In this case his total bet is

$$f_* = \frac{b_2 - 1}{2b_2} - \frac{b_1 - 1}{4}\left(\frac{1}{f_1} - \frac{1}{f_2}\right)\frac{f_2 - 1}{f_2 + 1}$$

which looks similar to the formula above for the Proebsting formulation, except that the sign is reversed on the second term and it is multiplied by an additional term. The two formulations are equivalent. In the original formulation, the bettor has 0.25 bet at 2 to 1 and 0.225 bet at 5 to 1. If he wins, he get 2.625 and if he loses he has 0.525. The second formulation makes clear that the change in behavior results from the mark-to-market loss the investor experiences when the new payout is offered. This is a natural way to think in finance, less natural to a gambler. In this interpretation, the infinite series of doubling payout does not ruin the Kelly bettor by enticing him to over bet, it extracts all his wealth through change beyond his control.

Bibliography

1. Kelly, J.L. (1956). A new interpretation of information rate, Bell System Techical Journal, 35, pp 917-926.
2. Pabrai, M. (2007). *The Dhandho Investor: The Low-Risk Value Method to High Returens,* Wiley, ISBN 978-0-470-04389-9.
3. Poundstone, Wlliam (2005) *Fortune's Formula: The untold story of the scientific betting system that beat the casinos and Wall Street,* New York: Hill and Wang, ISBN 0-8090-4637-7.
4. Thurp, E.O. (2008a). The Kelly criterion: Part I, Wlmott Magazine, May 2008.
5. Thurp, E.O. (2008b).Understanding the Kelly criterion: Part II, Wlmott Magazine, September 2008.

Chapter 36

Raven Paradox

THE PARADOX OWES ITS ORIGIN TO A NUMBER OF VERY DRASTIC POSITIONS ABOUT THE FORM IN WHICH SUCH UNIVERSAL STATEMENTS SHOULD BE EXPRESSED AND THE CONFIRMATION VALUE THE INDIRECT OBSERVATIONS SHOULD BE DEEMED TO HAVE.

The raven paradox, also named the paradox of black ravens, one of most renowned paradoxes arising from the confirmation procedures for universal statement of empirical theories. Its aim is to show that the paradox owes its origin to a number of very drastic positions about the form in which such universal statements should be expressed and the confirmation value the indirect observations should be deemed to have. More specifically, it is a paradox arising from the question of what constitutes evidence for a statement observing objects that are neither black nor ravens may formally increase the likelihood that all ravens are black, even though, intuitively, these observations are unrelated. This problem was proposed by the logician Carl Gustav Hempel (1940) illustrate a contradiction between inductive logic and intuition.

Traditionally, the Raven paradox is generated by the following two assumptions.

(a) Nicod Condition (NC): For any object 'a' and any predictive *F* and *G*, the proposition that has both *F* and *G* confirms the proposition that every *F* has *G*. A more formal version of

(NC) is the following claim
(Fa.Ga) confirms for all x *(Fx ⊃ Gx)*, for any individual term 'a' and any pair of predicates F and G.

(b) Equivalence Condition (EC): For any proposition H_1, H_2 and *E if E* confirms H_1 and H_1 is logically equivalent to H_2, then *E* confirms H_2.

From (NC) and (EC), we can deduce the following paradoxical conclusions (PC)

"The proposition that a is both non-black and non-raven (~Ba, ⊃ ~Ra), confirms the proposition that every raven is black, for all x (Rx Bx)"

The canonical derivation of (PC) from (EC) and (NC) proceeds as follows:

1) By (NC), (~Ba,~Ra) confirms for all x (*~Bx ⊃ Rx*).
2) By classical logic, for all x (*~Bx ⊃ ~Rx*) is equivalent to for all x (*Rx ⊃ Bx*).
3) By (1), (2) and (EC), (*~Ba .~Ra*) confirms for all x *Rx ⊃ Bx*.

But Hempel (1945) describes the paradox in terms of the hypothesis:

(A) All ravens are black

In st rict logical terms, via contraposition, this statement is equivalent to

(B) Everything that is not black is not a raven.

It should be clear that in all circumstances where (B) is true, (A) is also true and likewise, in all circumstances where (B) is false i.e., if a world is imagined in which something that was not black, yet was a raven, existed, (A) also false. This established logical equivalence.

Hemple (1945) and Goodman (1954) didn't view (PC) as paradoxical. Indeed, Hempel (1945) and Goodman (1954) viewed the argument about from (1) and (2) to (PC) as sound. So, as far as they are concerned; there is something misguided about whatever intuitions may have leaded some philosophers to see "paradox" here. As Hempel explains, one might be misled into thinking that (PC) is false by conflating (PC) with a different claim (PC*), a claim that is, in fact, false. Hempel warns us that

> ".....in the seemingly paradoxical cases of confirmation, we are often not judging the relation of the given evidence E alone to the hypothesis H ...... Instead, we tacitly introduce a comparison of H with a body of evidence which consists of E in conjunction with an additional amount of information we happen to have at our disposal."

One of the most popular proposed resolutions is to accept the conclusion that the observation of a green apple provides evidence that all ravens are black but to argue that the amount of confirmation provided is very small, due to the large discrepancy between the number of ravens and the number of non-black objects. According to this resolution, the conclusion appears paradoxical because we intuitively estimate the amount of evidence provided by the observation of a green apple to be zero, when it is in fact non-zero but extremely small. Good (1960) presented the argument is the best known and variations of the argument have been popular ever since Fitelson and Hawthorne (2006), although it had been presented by Alexander (1958) and early forms of the argument appeared as early as 1940. [Hosaisson - Lindenbaum (1940)]. Good (1960) argument involves calculation the weight of evidence provided by the observation of black raven or white shoes in favor of the hypothesis that all the ravens in a collection of objects are black. The weight of evidence is the logarithm of Bayes factor, which in this case is simply the factor by which the odds of the hypothesis changes when the observation is mode. The argument goes as follows:

> "......suppose that there are N objects that might be seen at any moment of which n are ravens and *r* are black and that the *N* objects each have probability $\frac{1}{N}$ of being seen. Let H_i be the hypotheses $H_1, H_2, \ldots\ldots\ldots, H_n$ are initially equip-probable. Then, if we happen to see a black raven, the Bayes factor in favor of H_0 is

$$\frac{\frac{n}{N}}{Average\left[\frac{n-1}{N},\frac{n-2}{n},\dots\dots\frac{1}{N}\right]}=\frac{2n}{n-1}$$

If the factor we see a white shoe is only

$$\frac{\frac{n-r}{n}}{Average\left[\frac{N-r-1}{N},\frac{N-r-2}{N}\dots\dots\dots\max\left(o,\frac{N-n-r}{N}\right)\right]}$$

$$=\frac{N-r}{\max\left[N-r-\frac{n}{2}-\frac{1}{2},\frac{1}{2}(N-r-1)\right]}$$

And this exceeds unity by only about

$\frac{n}{(2N-2r)}$ if (N–r) is large compared to n

Thus the weight of evidence provided by the sight of white shoe is positive, but is small if the number of ravens is known to be small compared to the number of non-black objects."

Many of proponents of this resolution and variants of it have been advocates of Bayesian probability and it is now commonly called the Bayesian solution, although, as Chihara (1987) observes. "There is no such thing as the Bayesian solution. There are many different 'solutions' that Bayesians have put forward using Bayesian techniques. Noteworthy approaches using Bayesian techniques include Earman (1992), Eells (1982), Gibson (1969) and Mackie (1963), who claims that his approach is more Bayesian than the so-called Bayesian solution of some paradox. Bayesian approaches which make use Carnap's theory of inductive inference include Humburg (1986), Maher (1999, 2004), Fitelson (1999, 2001, 2004), Vrancs (2002) introduced the term "Standard Bayesian Solution" to avoid confusion.

Carnup (1950, 1952, 1980) proposed various theories of "tautological confirmation" in terms of logical probability.

Maher (1999, 2004) has brought a carnapian approach to bear on the raven's paradox, with some very enlightening results. He accepts the paradoxical conclusion and refine it, a non-raven confirms that all ravens are black because

(i) the information that this object is not a raven removes the probability that this object is a counterexample to the generalization and

(ii) it reduces the probability that unobserved objects are ravens, thereby reducing the probability that they are counter example to the generalization.

In other word, to reach (ii), he appeals to Carnop's theory of inductive probability, which is, from the Bayesian point of view, a way of assigning prior probabilities which naturally implements induction. According to Carnap's theory, the posterior probability P $(Fa|E)$, that an object, *'a'* will have a predicate. 'F' after the evidence E has been observed is

$$P(Fa|E) = \frac{n_F + \lambda P(Fa)}{n + \lambda}$$

Where *P (Fa)* is the initial probability that a has the predicate *F*, n is the number of objects which have been examined, n_F is the number of examined objects which turned out to have the predicate *F* and λ is much larger than *n*, $P(Fa|E)$ will be very close to *P (Fa)* regardless of the fraction of observed objects which had the predicate F.

Using this carnapian approach, Maher (1999, 2004) identifies a proposition which we intuitively know to be false, but which we easily confuse with the paradoxical conclusion. The proposition in question is the proposition that observing non-ravens tells us about the color of ravens. While this in intuitively false and is also false according to Carnap's theory of induction, observing non-ravens causes us to reduce our estimate of the total number of ravens and thereby reduce the estimate of possible counter examples to the rule that all ravens are black. Hence, from the Bayesian-Carnapian point of view, the observation of non-raven does not tell us anything about the color of ravens, but it tells us about the prevalence of ravens and supports, all ravens are black, by reducing our estimate of the number of ravens which might not be black.

Fitelson (1999, 2001, 2004) examined the conditions under which the observation of a non-black non- raven provides less evidence than the at random, Ba is the proposition that the object is black and Ra is the proposition that the object is a raven, then condition

$$\frac{P\left(\overline{Ba}\middle|\overline{H}\right)}{P\left(\overline{Ra}\middle|\overline{H}\right)}-P\left(\overline{Ba}\middle|Ra\overline{H}\right)\geq P\left(Ba\middle|Ra\overline{H}\right)\frac{P\left(\overline{Ba}\middle|H\right)}{P\left(Ra\middle|H\right)}$$

is sufficient for the observation of non-black non-raven to provide less evidence than observation of a black raven. Here, a line over a proposition indicates the logical negation of that proposition. This condition does not tell us how large the difference in the evidence provide is, but a later calculation in same paper shows that the weight of evidence provided by a black raven exceeds that provided by a non-black non-raven by about

$$-\log P\left(Ba\middle|Ra\overline{H}\right)$$

This is equal to the amount of additional information which is provided when a raven of unknown color is discovered to be black, given the hypothesis that not all ravens are black. He explain that

> "Under normal circumstances, $p = P\left(Ba\middle|Ra\overline{H}\right)$ may be somewhere around 0.9or 0.95, so $\frac{1}{p}$ is somewhere around 1.11 or 1.05. Thus, it may appear that a single instance of a black raven does not yield much more support than would a non-black non-raven. However, under plausible conditions it can be shown that a sequence of n instances yields a ratio

of likelihood ratios on the order of $\left(\frac{1}{p}\right)^n$, which blows up significantly for large n".

He point out that his analysis is completely consistent with the supposition that a non-black non-raven provides an extremely small amount of evidence although they do not attempt to prove it, the merely calculate the difference between the amount of evidence that a black raven provides and the amount of evidence that a non-black non-raven provides.

Bibliography

1. Alexander, H.G. (1958). The Paradox of Confirmation, British Journal for the Philosophy of Science, 9, pp 227-233.
2. Armstrony, D. (1978). A Theory of Universals, Cambridge University Press, Cambridge.
3. Carnap, R. (1950). Logical Foundation of Probability, 2ed, Chicago University Press, Chicago.
4. Carnap, R. (1952). The Continuum of Inductive Method, University of Chicago Press, Chicago.
5. Carnap, R. (1980). A Basic System of Inductive Logic, Part II, in Richard C. Jeffrey (ed.), Studies in Inductive Logic and Probability, 2, University of California Press, Berkeley.
6. Chihara, C. (1981). Quine and the Conformational Paradox in Peter A. French, Theodore E. Uehling Jr. and Howard k. Wettstein (eds.), Midwest Studies in philosophy, 6, The Foundations of Analytic Philosophy, University of Minnesota Press, Minneapolis, pp 425-452.
7. Earman, J. (1992). Bayes or Bust? A Critical Examination of Bayesian Confirmation Theory, MA: MIT Press, Cambridge.
8. Eells, E. (1982). Rational Decision and Causality, Cambridge University Press, New York.
9. Fitelson, B. (1999). The Plurality of Bayesian Measure of Confirmation and the Problem of Measure Sensitivity, Philosophy of Science, 66, pp 362-378.
10. Fitelson, B. (2001). Studies in Bayesian Confirmation Theory, Ph.D. Thesis, University of Wisconsin.
11. Fitelson, B. (2004). Inductive Logic forthcoming in the

Philosophy of Science: An Encyclopedia, eds. Pfeifer, J. and S.Sarkar, Routledge Pree.

12. Gaifman, Haim (1979). Subjective Probability, Natural Predicates and Hempel's Ravens, Erkenntnis, pp 105-147.
13. Gemes, Ken (1999). Carnap- Confiematiotion, Content Cutting and real confirmation, Yale, unpublished essay.
14. Gibson, L. (1969). On Ravens and Relevance and a Likelihood Solution of the Paradox of Confirmation, British Journal for the Philosophy of Science, 20, pp 75-80.
15. Good, I.J. (1960). The Paradox of Confirmation, British Journal for the Philosophy of Science, 11, pp 145-149.
16. Good, I.J. (1961). The Paradox of Confirmation II, British Journal for the Philosophy of Science, 12, pp 63-64.
17. Good, I.J. (1967). The White Shoe is a Red Herring, British Journal for the Philosophy of Science,17, pp 322.
18. Good, I.J. (1968). The White Shoe qua Herring is Pink, British Journal for the Philosophy of Science,19, pp 156-157.
19. Goodman, N. (1954). Fact, fiction and Forecast, Athlone Press, London.
20. Hempel, C.G. (1945). Studies in the logic of confirmation I and II Mind, 54.
21. Hempel, C.G. (1967). The White Shoe: No Red Herring, British Journal for the Philosophy of Science,18, pp 239-240.
22. Hosiasson-Lindenbaum, Janina (1940). On Confirmation, Journal of Symbolic logic, 5, pp 133-148.
23. Humburg J. (1986). The solution of Hempel's Raven Paradox in Rudolf Carnap's system of inductive logic, Erkenntnis, 24, pp 57-72.
24. Maher, P. (1999). Inductive logic and Ravens Paradox, Philosophy of Science, 66, pp 50-70.
25. Maher, P. (2004). Probability capture the logic of scientific confirmation in contemporary debates in the Philosophy of Science, ed., Christopher Hitchcock, Oxford: Blackwell, pp 69-93.
26. Woodward, J. (1985). Critical Review: Harwich in the Ravens, Project ability and Induction, Philosophical Studies, 47, pp409-428.

Chapter 37

Relevance Paradox

SOMETIMES RELEVANT INFORMATION IS NOT SOUGHT OUT BECAUSE ITS RELEVANCE ONLY BECOMES CLEAR AFTER THE INFORMATION IS AVAILABLE.

The relevance paradox describes an attempt to gather information relevant to decision, which fails because the elimination of information perceived as distracting or unnecessary and thus detrimental to making an optimal decision, can also inadvertently exclude, information that is actually crucial. In other words, the relevance paradox occurs where individual, professionals or groups of professionals are unaware of certain essential information which would guide them to better decisions, and help avoid inevitable, unintended and undesirable consequences but lacking the information they don't see it relevance so don't seek it. Affected persons will only seek the information and advice they think they need as opposed to what they actually need to meet their own or the organization's goals.

In many cases in which action or decision is required, it is obvious what information relevant to the matter at hand may be lacking, a military attack may not have maps so reconnaissance is undertaken, and engineering project may not have ground condition details and these will be ascertained a public health program will require a survey of which illnesses are prevalent and so on. The civil engineers were victims of the relevance paradox to the extent. They only thought they needed to know about concrete, water flows etc. Not how to restrict velocities to prevent the snail species which carried the disease form multiply. The another example include the NASA engineers who having spent a fortune on unsuccessfully developing the complex sliding and articulating inside knee joint needed for space suite eventually went to the Tower of London and ruefully copied the armor of Henry the Eighth. Or the astonishing and not widely appreciated fact that United Kingdom wastes heat from power stations equal to the entire import of natural gas for heating, which is an amazing example of the relevance paradox and hierarchical incompetence of particular relevance 30 years on now that fears are growing about inadequate reserves of natural gas.

Another example from civil engineering unwittingly caused a massive increase in the debilitating water borne infection schistosomiase (bilharzias) for locals as a result of irrigation schemes that lacked simple low-cost countermeasures built in simply because they had no knowledge of these countermeasures. Yet at the same time, the United Nations had already published guidelines explaining cheap countermeasures and how they could be built into the design of the irrigation schemes, matters as simple as keeping velocities above a certain level to prevent the disease vector from attaching to the conduits. The civil engineers were victims of the relevance paradox because they thought they only needed to know about engineering issues such as concrete and water flows, not how to control flow velocities to prevent the snail species that carried the disease from multiplying, so they failed to seek this information. The relevance, paradox can usually does apply to all professional groups and individuals in numerous ways. While there are many examples of willful ignorance, there are also many cases where people do not look outside the paradigms they are operating in and thus fail to see the long term consequences.

However, in many significant instances across a wide range of areas, even when relevant information is readily available, the decision makers are not aware of its relevance because they lack the information which would make its relevance clear. As a result, they do not attempt to look for it. These decision makers will seek only the information and advice they believe is the minimum amount required as opposed to what they actually need to fully meet their goals. An analogy would be that of near-sighted person who is unaware of their condition. They would not perceive any benefit in getting the glasses that they need to improve their sight, until they tried wearing the glasses. Such a situation could be resolved by a third party, aware of its relevance, recommending an eye test. The relevance paradox is cited as a cause of the increase in disease in developing countries even while more money is being spent on them. Relevance paradoxes occur because of implementation of projects without awareness of the social or individual tacit knowledge within a target community. The understanding of the individual and the social tacit knowledge in a given community, which is a function of knowledge emergence, is the foundation of effectiveness in the leadership practice.

Bibliography

1. Andrews, David (1984). *The IRG Solution - Hierarchical Incompetence and how to overcome it.* London: Souvenir Press. pp. 200-220. *ISBN 0-285-62662-0.*
2. http://*therelevanceparadox.blogspot.in/*
3. https://*andrewlainton.wordpress.com/tag/relevance-paradox/*
4. http://*www.claverton-energy.com/energy-experts-online/relevance-paradox*

Chapter 38

Simpson's Paradox

A TREND THAT APPEARS IN DIFFERENT GROUPS OF DATA DISAPPEARS WHEN THESE GROUPS ARE COMBINED, AND THE REVERSE TREND APPEARS FOR THE AGGREGATE DATA.

In statistics, Simpson's paradox or Yule-Simpson effect is a paradox in which a trend that appears in different groups of data disappears when these groups are combined and the reverse trend appears for the aggregate data. Simpson paradox introduced by Blyth (1972) and first encountered by Pearson in 1899 [Aldrich (1995)], refers to the phenomenon where by an event *C* increases the probability of *E* in a given population *p* and at the same time, decreases the probability of E in every subpopulation of *p*. In other words, if *F* and F' are two complementary properties describing two subpopulations, we might well encounter the inequalities

$$P(E|C) > P(E|C') \tag{1}$$

$$P(E|C,F) < P(|E|C',F) \tag{2}$$

$$P(E|C,F') < P(E|C',F') \tag{3}$$

Although such order reversal might not surprise students of probability, it is paradoxical when given causal interpretation. For example, if we associate *C* with taking a certain drug, *E* with recovery and *F* with being a female then under the causal interpretation of (2) and (3), the drug seems to be harmful to both males and females yet beneficial to the population as a whole intuition deems such a result impossible and correctly so.

In other words, consider two populations for which the overall rate r of occurrence of some phenomenon in population A is greater than the corresponding rate R in population B. Suppose that each of the two populations is composed of the same two categories C_1 and C_2 and the rates of occurrence of the population *B* are R_1 and R_2. If $r_1 < R_1$ and $r_2 < R_2$, despite the fact that $r > R$ then Simpson's paradox is said to have occurred. For example, which consist of the batting averages of two baseball players.

	Player A			**Player B**		
	Time at bat	Hits	Average	Time at bat	Hits	Average
Against right handed pitcher (C_1)	202	45	0.223 (=r_1)	250	58	0.232 (= R_1)
Against left handed pitcher (C_2)	250	71	0.284 (= r_2)	108	32	0.296 (= R_2)
Overall	452	116	0.257 (= r)	358	90	0.251 (= R)

The key to the paradox is the differential weight of the rates for C_1 and C_2 when the overall rate for each population is computed. Since

$$r = w_1 r_1 + w_2 r_2 \text{ and } R = W_1 R_1 + W_2 R_2$$

where $w_1 + w_2 = 1$ and $W_1 + W_2 = 1$

there is no necessary ordering of r and R no matter what r_1, r_2, R_1 and R_2 are. In this example, Player A batted against right-handed pitchers 202 out of 452 times at bat or 44.7% of the time, w_1 = 0.447 and w_2 = 0.553, whereas Player B batted against right-handed pitcher 250 out of 358 times at bat or 69.8% of the time, W_1 = 0.698 and W_2 = 0.302. This discrepancy in relative weights was large enough to switch the rank-order of the batting averages of the two players. So Simpson's paradox can occur.

Apart from the numerical representation of Simpson's paradox, there is another way to illustrated that is using the two-dimensional vector space. $\vec{A}$ success rate of $\frac{p}{q}$ can be represented by a vector $\vec{A}$ = (q, p) with slope of $\frac{p}{q}$. If two rates $\frac{p_1}{q_1}$ and $\frac{p_2}{p_2}$ are combined the results can be represented by the vectors (q_1, p_1) and (q_2, p_2), which according to the parallelogram rule is the vector $(q_1 + q_2, p_1 + p_2)$, with slope $\frac{p_1 + p_2}{q_1 + q_2}$

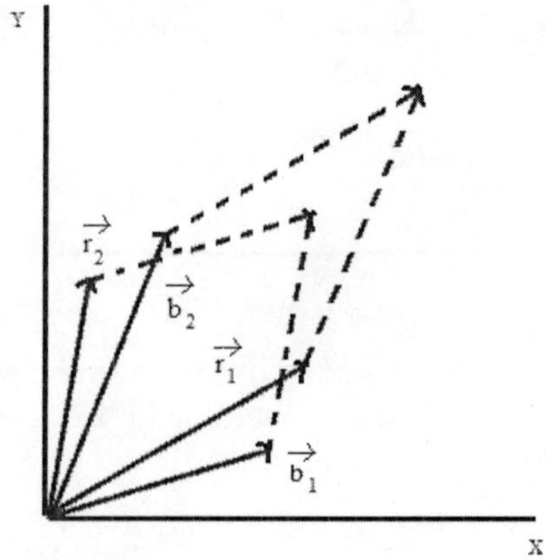

Simpson's paradox says that even if vector $\vec{b_1}$ has smaller slope than another vector $\vec{r_2}$ and $\vec{b}_2$ has smaller slope than $\vec{r_2}$, sum of the two vectors $\vec{b_1}+\vec{b}_2$ can still have a larger slope than the sum of the two vector $\vec{r}_1+\vec{r}_2$.

After understanding the statistical agony behind this paradox the aim left is to solve it. In the case of Simpson's paradox, we have a clash

(i) between the assumption that causal relationships are governed by the laws of probability calculus.

(ii) the set of implicit assumptions that drive our causal intuitions.

The first assumption tells us that the three inequalities in (1) to (3) are consistent. To resolve the paradox we must either

(a) show that our causal intuition is misleading or incoherent or

(b) deny the premise that causal relationships are governed by the laws of standard probability calculus.

Here we use second option, is that causality is governed by its own logic and that this logic requires a major extension of probability calculus. This still behooves us to explicate the logic that governs our causal intuition and to show logic of the do operator is perfectly suitable for this purpose. Let us first translate the statement that our drug C has harmful effect on both male and females into formal statements in causal calculus.

$$P\left(E\middle|do(C),F\right)<P\left(E\middle|do(C'),F'\right) \tag{4}$$

$$P\left(E\middle|do(C),F'\right)<P\left(E\middle|do(C'),F'\right) \tag{5}$$

We need to demonstrate that C must be harmful to the population at large i.e. the inequality.

$$P\left(E\middle|do(C)\right)<P\left(E\middle|do(C')\right) \tag{6}$$

must be shown to be inconsistent with what we know

about drugs and gender. For this, use the Sure-Thing Principle which states that "an action C that increases the probability of an event *E* in each subpopulation must also change the distribution of the subpopulations. Here we need to prove that the reversal in the inequalities of (4), (6) is inconsistent with the assumption that drugs have effect on gender.

$$P(E|do(C)) = P(F|do(C')) = P(F) \tag{7}$$

Now expanding the term $P(E|do(C))$, we get

$$P(E|do(C)) = P(E|do(C)F)P(F|do(C)) + P(E|do(C),F')P(F'|do(C))$$

$$= P(E|do(C),F)P(F) + P(E|do(C),F')P(F')) \tag{8}$$

Similarly for do(C'), we get

$$P(E|do(C')) = P(E|do(C'),F)P(F) + P(E|do(C'),F')P(F') \tag{9}$$

Since every term on the R.H.S of (8) is smaller than corresponding term in (9), we conclude that

$$P(E|do(C)) < P(E|do(C'))$$

Here our intuitive logic tells us that it is perfectly consistent to find a drug satisfying the three inequality of (4) to (6) and that it would be inappropriate to adjust for *F*. Fortunately, the mathematical tools that have been developed in the past years permit a more systematic and friendly resolution of Simpson's paradox but we adopt the above method to resolved it.

Bibliography

1. Aldrich, J. (1995) Correlations genuine and spurious in Pearson and Yule, Statistical Science, 10,pp127-133.
2. Blyth, C.R. (1972). On Simpson's paradox and the Sure Thing Principle, Journal of the American Statistical Association, 67,pp 364-366.
3. Cartwright, N. (1983). *How the laws of Physics Lie,* Clarendor

Press Oxford.

4. Cohen, M.R. and Nagel, E. (1934). *An International to Logic* and the Scientific Method, Harcourt, Brace and Company, New York.
5. Dong, J. (1998). Simpson's paradox: Encyclopedia of Biostatistics, Wiley, New York, pp 4108-4110.
6. Everitt, B. (1995). Simpson Paradox, The Cambridge Dictionary of Statistics in the Medical Science, pp 337. Cambridge University Press, New York.
7. Jery, K.(2001). Proofs without Words: Simpson Paradox, Mathematics Magazine, 74(5).
8. Simpson, E.H. (1951). The Interpretation of Interaction in Contingency Table. Journal of the Royal Statistical Society, Series B, 13, pp 238-241.

Chapter 39

Sleeping Beauty Problem

A PROBABILITY PROBLEM THAT CAN BE CORRECTLY ANSWERED AS ONE HALF OR ONE THIRD DEPENDING ON HOW THE QUESTION IS APPROACHED.

The Sleeping beauty problem is a puzzle in probability theory and formal epistemology in which an ideally rational epistemic agent is to be woken once or twice according to the toss of a coin and asked her degree of belief for the coin having come up heads. It is a variation of some very similar problems of imperfect recall that have, been discussed for some time in the game theoretic literature. It was named by Robert Stalnaker, who had learnt about similar cases from Zuboff (1990). The problem was brought to the attention of the philosophical community through an exchange between Adam Elga, who argued for the answer $\frac{1}{3}$ and Devid Lewis, who defended the $\frac{1}{2}$ view.

In general, sleeping beauty is an example of a problem involving self-locating beliefs i.e. beliefs that an agent or a temporal part of an agent, might have about its own location. An agent part that knew exactly which possible world is actual can still be ignorant about its own location in the world. That can happen if the world contains two or more agent parts whose evidential states are subjectively indistinguishable. These agent parts would then be unable to determine with certainty their own philosophical location. Ever if beauty knew that the outcome of coin toss would be tail, she could not know whether it was currently Monday or Tuesday in the original Sleeping Beauty problem i.e.

> "On Sunday, Beauty is put to sleep. She is awakened once on Monday and put to sleep again after being administered a memory-erasing drug that causes her to forget her wakening. A fair coin is tossed. If and only if the coin fall tail, Beauty is awaking again on Tuesday. She knows all this when she awakes on Monday, what should her credence be that the coin will fall heads?"

Another way of expressing this by saying that the agent parts would be ignorant about which centered possible would they are in even though they know which possible would they are in. This problem is but one piece of the larger puzzle of how to relate indexical to non-indexical information in our reasoning. If one studies the problem in isolation from this wider context, one rists coming up with answers and principles that do not fit with other parts of the puzzle. Here we argue that the correct answer is $\frac{1}{3}$.

For this, imagine the experiment is repeated many times. Then in the long run, about $\frac{1}{3}$ of the waking's would be Heads-waking's, waking's that happen on tails in which the coin lands Heads, so on any particular waking, you should credence $\frac{1}{3}$ that waking is a Heads-waking and hence credence $\frac{1}{3}$ in the coin's landing Heads on that trial. Suppose that the first waking happens on Monday and that the second waking happens on Tuesday. Then when you wake up, you are in one of the three.

H_1: HEADS and it is Monday

T_1: TAILS and it is Monday

T_2: TAILS and it is Tuesday

Notice that the difference between your being in T_1 and your being in T_2 is not a difference in which possible world is actual, but rather a difference in your temporal location within the world. In a more technical treatment we might adopt a framework similar to the one suggested in Lewise (1983) according to which the elementary alternatives over which your belief is divided are not possible worlds, but rather centered possible world (possible worlds each of which is equipped with a designated individual and time). In such a framework H_1, T_1 and T_2 would be represented by appropriate sets of centered worlds.

Let P be the credence function (believe function) i.e. you believe that you ought to have upon first awakening. Upon first awakening, you are certain of the following, you are in predicament H_1 iff the outcome of the coin toss is Heads. Therefore, calculating $P(H_1)$ is sufficient to solve the Sleeping Beauty problem.

If you were to learn that the toss outcome is Tails; that would amount to your learning that you are in either T_1 or T_2. Since being in T_1 is subjectively just like being in T_2 and since exactly the same propositions are true whether you are in T_1 or T_2 even a highly restricted principle of indifference yields that you ought then to have equal credence in each. But your credence that you are in T_1 after learing that the toss outcome is tails, ought to be the same as the conditional credence $P(T_1 \, or \, T_2)$ and likewise for T_2. So

$$P(T_1 | T_1 \, or \, T_2) = P(T_2 | T_1 \, or \, T_2)$$

and hence $P(T_1) = P(T_2)$

Now, if you were to learn that it is Monday, then would amount to your learning that you are in either H_1 or T_1. Your credence that you are in H_1 would then be your credence that a fair coin, soon to be tossed will land Heads. It is irrelevant that you will be awakened on the following day iff the coin lands Tails, in this circumstance, your credence that the coin will land Heads ought to be $\frac{1}{2}$. But your credence that the coin will land Heads, after learning that it is Monday, ought to be the same as the conditional credence $P(H_1|H_1 \text{ or } T_1)$) or T_1). So

$$P(H_1|H_1 \text{ or } T_1) = \frac{1}{2}$$

and hence $P(H_1) = P(T_2)$

Combining results, we have that

$$P(H_1) = P(T_1) = P(T_2)$$

Since there are credence's probabilities so their sum is 1, therefore P $(H_1) = \frac{1}{3}$.

If we set $P(H_1) = P(T_1) = P(T_2)$, it follows that Beauty should, upon awakening, assign P(HEADS)= $\frac{1}{3}$ and P(TAILS)= $\frac{2}{3}$, the only relevant difference between hypotheses. Since the structure of the situation does not depend on the particular numbers involved, we can test our intuitions by considering a more extreme version of the problem, like Extreme Sleeping Beauty.

> "In this problem, except that here, if coin falls tails, Beauty will be awakened on a million subsequent days. As before she will be given an amnesia drug each time she is put to sleep that make her to forget get any previous awakenings. When she awakes on Monday, what should be her credence in HEADS?"

The protagonist of the $\frac{1}{2}$ view will maintain that Beauty, upon awakening, should retain her credence of $\frac{1}{2}$ in HEADS, but also that, upon being informed that it is Monday she should become extremely confident in HEADS.

$$P(\text{HEADS}) = \frac{1000001}{1000002} \approx 0.99999$$

This consequence is itself quite implausible. Now reasoning exactly parallel to that which Elga (2000) used to support the $\frac{1}{3}$ view in the original version, we obtain

$$P(H_1) = P(T_1) = P(T_2) = \ldots\ldots\ldots\ldots = P(T_{1000001})$$

Therefore Beauty should assign

$$p(\text{HEADS}) = \frac{1}{1000002}$$

This consequence in Extreme Sleeping Beauty is counterintuitive. It seems like a rather excessive confidence in the proposition that a fair coin, yet to be tossed, will fall tails.

Bibliography

1. Arntzenius, F. (2002). Reflections on Sleeping Beauty, Analysis, 62(1), pp 53-62.
2. Bostron, N. (2001). The Doomsday argument, Adam & Eve, UN++, and Quantum Joe., Synthese, 127(3), pp 359-387.
3. Bostron, N. (2002). Self-locating Belief in Big Worlds: Cosmology's Missing Link to observation, Journal of Philosophy, 99(12), pp 607-623.
4. Bostron, N. (2003). The Mysteries of Self-Locationg Belief and Anthropic Reasoning. Harvard Review of Philosophy, 11, pp 59-74.
5. Bostron, N. (2007). Sleeping Beauty and Self-location: A Hybrid Model, Synthese, 157(1), pp 59-78.
6. Elga, A (2000). Self-location Belief and the Sleeping Beauty and the Sleeping Beauty problem, Analysis, 60(2), pp 143-147.

7. Lewis, D. (2001). Sleeping Beauty: Reply to Elga, Analysis, 61, pp 171-176.
8. Weintraub, R. (2004). Sleeping Beauty: A Simple Solution, Analysis, 64(1), pp 8-10.

Chapter 40

Stein's Paradox

IT IS THE PHENOMENON THAT WHEN THREE OR MORE PARAMETERS ARE ESTIMATED SIMULTANEOUSLY, THERE EXIST COMBINED ESTIMATORS MORE ACCURATE ON AVERAGE THAN ANY METHOD THAT HANDLES, THE PARAMETERS SEPARATELY.

The best guess about the future is usually obtained by computing the average of past events. Stein (1956) defines it as circumstances in which there are estimators better than the arithmetic average. In decision theory and estimation theory, Stein's paradox is the phenomenon that when three or more parameters are estimated simultaneously, there exist combined estimators more accurate on average than any method that handles, the parameters separately. An intuitive explanation is that optimizing for the mean-squared error of a combined estimator is not the same as optimizing for the errors of separate estimators of the individual parameters. In practical terms, if the combined error is in fact of interest, then a combined estimator should be used, even if the underlying parameters are independent, this occurs in channel estimation in telecommunications, for different factors affecting overall channel performance. On the other hand, if one is instead in estimating an individual parameter, then using a combined estimator does not help and is in fact worse.

Perhaps the most surprising result in statistics arises in remarkably simple estimation problem. Let θ be a vector consisting of $n \geq 3$ unknown parameters. To estimate this parameter, a single measurement x_i is performed for each parameter θ_i, resulting in a vector X of length n. Suppose the measurement are independent, Gaussian random variable, with mean θ and variance one.

$$X \sim N(\theta, I)$$

Thus, each parameter is estimated using a single noisy measurement and each measurement is equally inaccurate. Under such conditions, it is most intuitive to use each measurement as an estimate of its corresponding parameter. This is known as the ordinary decision rule and can be written as

$$\hat{\theta} = x$$

The quality of such an estimator is measured by its risk function. A commonly used risk function is the squared error, defined as

$$E\left[\left\|\theta - \hat{\theta}\right\|^2\right]$$

Surprisingly, it turns out that the ordinary estimator proposed above is suboptimal in terms of mean squared error when $n \geq 3$. In other words, there exist alternative estimators which always achieve lower mean squared error, no matter what the value of θ is. For given θ one could obviously defined a perfect estimator which is always just θ, but this estimator would be bad for other values of θ. The estimator of Stein's paradox are, for a given θ, better than X for some values of X but necessarily worse for others, except perhaps for one particular θ vector, for which the new estimate is always better than X.

More strictly, and estimator $\hat{\theta}_1$ is said to dominate another estimator $\hat{\theta}_2$ if, for all values of θ, the risk of $\hat{\theta}_1$ is lower than, or equal to, the risk of $\hat{\theta}_2$ and if the inequality is strict for some θ. An estimator is said to be admissible if no other estimator dominates it, otherwise it is inadmissible. Thus, Stein's paradox can be simply stated as

> "The ordinary decision rule for estimating the mean of multivariate Gaussian distribution is inadmissible under mean squared error risk in dimension at least three."

The risk function of the decision rule d(x) = x is

$$R(\theta,d) = E_\theta\left[|\theta - x|^2\right]$$

$$= \int(\theta - x)'(\theta - x)\left(\frac{1}{2\pi}\right)^{n/2} e^{-\frac{1}{2}(\theta - x)'(\theta - x)} m\,(dx)$$

$$= n$$

Now. consider the decision rule

$$d^*(x) = x - \frac{a}{|x|^2}x$$

where α = n–2. We will show that d* is better decision rule than d. The risk function is

$$R(\theta,d) = \left[|\theta - X + \frac{a}{|x|^2}X|^2\right]$$

$$= E_\theta\left[|\theta - X|^2 + 2(\theta - X)'\frac{a}{|X|^2}X + \frac{a^2}{|X|^4}|X|^2\right]$$

We may simplify the middle term by considering a general well-behaved function

$$h : x \to h(x) \in R$$

and using integration by part, for $1 \leq i \leq n$, for any continuously differentiable h growing sufficiently slower for large x_i we have

$$E_\theta\left(\left(\theta - X_i\right)h(X)\middle|X_j = x_j\left(j \neq i\right)\right)$$

$$= \int\left(\theta_i - x_i\right)h(x)\left(\frac{1}{2\pi}\right)^{n/2} e^{-\frac{1}{2}(\theta-x)'(\theta-x)} m\left(dx_i\right)$$

$$= \left[h(x)\left(\frac{1}{2\pi}\right)^{n/2} e^{-\frac{1}{2}(\theta-x)'(\theta-x)}\right]_{xi=-\infty}^{\infty} - \int\frac{\partial h}{\partial x_i}(x)\left(\frac{1}{2\pi}\right)^{n/2} e^{-\frac{1}{2}(\theta-x)'(\theta-x)} m\left(dx_i\right)$$

$$= -E_\theta\left(\frac{\partial h}{\partial x_i}(X)\middle|X_j = X_j\left(j \neq i\right)\right)$$

therefore

$$E_\theta\left[\left(\theta - x_i\right)h(X)\right] = -E_\theta\left[\frac{\partial h}{\partial x_i}(X)\right]\Bigg]$$

This is known as the Stein's lemma.

Now choose

$$h(x) = \frac{x_i}{|x|^2}$$

If h met the well-behaved condition, we would have

$$\frac{\partial h}{\partial x_i} = \frac{1}{|x|^2} - \frac{2x_i^2}{|x|^4}$$

And so

$$E_\theta\left[\frac{(\theta - X)'X}{|X|^2}\right] = \sum_{i=1}^{n} E_\theta\left[\left(\theta_i - x_i\right)\frac{X_i}{|X|^2}\right]$$

$$= -\sum_{i=1}^{n} E_\theta\left[\frac{1}{|X|^2} - \frac{2X_i^2}{|X|^4}\right]$$

$$= -(n-2)E_\theta\left[\frac{1}{|X|^2}\right]$$

Then returning to the risk function of d*

$$R(\theta, d^*) = n - 2\alpha(n-2)E_\theta\left[\frac{1}{|X|^2}\right] + \alpha^2 E_\theta\left[\frac{1}{|X|^2}\right]$$

This quadratic in α is minimized at α = n - 2, giving

$$R(\theta, d^*) = R(\theta, d) - (n-2)^2 E_\theta\left[\frac{1}{|X|^2}\right]$$

which of course satisfies

$$R(\theta, d^*) < R(\theta, d)$$

makingd an inadmissible decision rule. It remains to justify the use of

$$h(x) = \frac{X}{|X|^2}$$

This function is not continuously differentiable since it is singular at x=0. However the function

$$h(x) = \frac{X}{\in + |X|^2}$$

is continuously differentiable and after following the algebra through and letting $\in \to 0$ one obtain the same result.

Stein's paradox is surprising; since the ordinary rule is intuitive and commonly used. In fact, numerous methods for estimator construction, including maximum likelihood estimation, best linear unbiased estimation, least square estimation and optimal equivalent estimation, all result in the ordinary estimator. Yet, as discussed earlier, this estimator is suboptimal. To demonstrate the unintuitive nature of Stein's paradox, consider the following real-world example. Suppose we are to estimate three unrelated parameters, such as the wheat yield for 2013 in India, the number of spectators at the Wimbledon tennis tournament in 2013 and the weight of a randomly chosen candy bar from the supermarket. Suppose we have independent Gaussian measurements of each of these quantities. Stein's paradox now tells us that we can get better estimate for the vector of three parameters by simultaneously using the three unrelated measurements. At first sight it appears that somehow we get a better estimator for wheat yield by measuring some other unrelated statics such as the number of spectators at Wimbledon and the weight of a candy bar. This is of course absurd, we have not, obtained a better estimator for wheat yield by itself, but we have produced an estimator the vector of the means of all three random variable, which has a reduced total risk. This occurs because the cost of a bad estimate in one component of the vector is compensated by better estimate in another component. Also, a specific set of the three estimated mean values obtained with the new estimator will not necessarily be better than the ordinary set. It is only on average that the new estimator is better.

Samworth (2005), baseball data example is another good example of this paradox which supplies convenient material for illustrating the workings of this paradox. The following table gives the baseball batting average of $p = 9$ baseball players, all of whom were active in 1990. For $i= 1,2,3,\ldots\ldots,p$, let n_i and z_i respectively denote the number of times at bat and batting average of ith player during the 1990 season. The following table showing number of times at bat n_i, batting average z_i in 1990 and career batting average π_i of $p = 9$ baseball players

Player	n_i	z_i	π_i
Baines	415	0.284	0.289
Barfield	476	0.246	0.256
Bell	583	0.254	0.265
Biggio	555	0.276	0.287
Bonds	519	0.301	0.297
Bonilla	625	0.280	0.279
Brett	544	0.329	0.305

Brooks Jr.	568	0.266	0.269
Browne	513	0.267	0.271

Here π_i denotes the player's true batting average, taken to be his career batting average. Each player had at least 3000 at bats in his career. We consider the model where $z_1, z_2, \ldots\ldots, z_p$ are independent, with $z_i \sim n_i^{-1}$ Bin (n_i, π_i). We make the transformation

$$X_i = \sqrt{n_i} S\text{in}^{-1}(2_{Zi} - 1)$$

and let $z_i = \sqrt{n_i} Sin^{-1}(2\pi_i - 1)$, which means that X_i is approximately distributed as $N(\theta_i, 1)$. A heuristic argument to justify this that by Taylor expansion applied to the function

$$g(x) = \sqrt{n_i} Sin^{-1}(2x - 1)$$

We have

$$X_i - \theta_i = g(z_i) - g(\pi_i)$$

$$\approx g'(\pi_i)(z_i - \pi_i)$$

$$= \frac{\sqrt{n_i}(z_i - \pi_i)}{\pi_i(1 - \pi_i)}$$

and this latter expression has an approximate N(0,1) distribution when n_i is large by the central limit theorem. In fact, since $\min_i n_i \geq 400$ an exact calculation given that the variance of each X_i is between 1 and 1.005 for $\pi_i \in [0.2, 0.8]$. For our prior guess

$$\theta_0 = (\theta_{0.1}, \theta_{0,2} \ldots \theta_o P)'$$

We take $\theta_{0,1} = \sqrt{\bar{n}}\, Sin^{-1}(2\pi_0 - 1)$

with $\pi_0 = 0.275$ and $\bar{n} = \frac{1}{p}\sum_{i=1}^{p} n_i$

We find that $\|X-\theta\|^2 = 2.56$,, some what below its expected value of around 9, this observation is only around 1.5 standard deviations away from its mean. On the other hand, $\|\theta-\hat{\theta}\|^2 = 1.5$, so Stein estimation does provide an improvement in this case. Letting $\pi = (\pi_1, \pi_2, \ldots\ldots\ldots, \pi_p)$ and recalling that θ is a function of π, the usual 95% confidence set for π

$$\{\pi \in [0,1]^p : \|X-\theta\|^2 \le 16.9\}$$

Bibliography

1. Efron, B and Morris, C.N. (1997). Stein's paradox in statistics, Scintific American, 236(5), pp 119-127.
2. James, W. and Stein, C. (1961). Estimation with quadratic loss, Proc. Fourth Berkeley Symposium, California University Press, pp 361-380.
3. Lehmann, E.L. (1983). *Theory of Pint Estimation*, Wiley, New York.
4. Samworth, R. (2005). Small confidence sets for the mean of spherically symmetric distribution, Journal of Royal Statistics Society, Series B, 67, pp 343-361.
5. Shao, P.Y.S. and Strawderman, W.E. (1994). Improving on the James stein positive past estimator, Annuals of Statistics, 22, pp 1517-1538.
6. Stein, C. (1956). Inadmissibility of the usual estimator of the means of a multivariate normal distribution, Proc.-Third Berkeley Symposium, California University Press, pp 197-206.

Chapter 41

Two-Envelope Problem

YOU ARE GIVEN TWO INDISTINGUISHABLE ENVELOPES, EACH OF WHICH CONTAINS A POSITIVE SUM OF MONEY. ONE ENVELOPE CONTAINS TWICE AS MUCH AS THE OTHER. YOU MAY PICK ONE ENVELOPE AND KEEP WHATEVER AMOUNT IT CONTAINS. YOU PICK ONE ENVELOPE AT RANDOM BUT BEFORE YOU OPEN IT YOU ARE GIVEN THE CHANCE TO TAKE THE OTHER ENVELOPE INSTEAD.

The two envelopes problem, also known as the Exchage paradox, is a brain teaser, puzzle or paradox in logic, philosophy, special interest in decision theory and for the Bayesian interpretation of probability theory. Historically, Christensen and Utts (1992) addressed the Two-envelops problem or Exchage problem. Many letters and articles have been written on this problem, mostly in Philosophy journals, where according to us, the basic inference problem, is still overlooked in most of them. Also according to us, the Bayesian analysis one has to identify a prior implied by the "paradoxical reasoning in the original setting. The problem of two-envelope paradox referred also as the problem of wallets and is as follows,

> "A Benefactor places a certain amount of money in one envelope and places twice that quantity in another envelope. The initial amount and the identification of the envelope

where the larger quantity was placed are unknown to you. Then, the Benefactor selects randomly (equal probability) one envelope and hands it to you and the contents is yours."

The mental reasoning to get the paradoxical result is, if X is inside your envelope, then even before opening it, you realize that the contents of the other envelope, call it Y, will be either $\frac{X}{2}$ or 2X, which due to the random selection yields an "expected value", say $E_{X,}$ for the contents of the other envelope, we have

$$Ex(Y) = \frac{1}{2} \times \frac{X}{2} + \frac{1}{2} \times 2X = \frac{5}{4} X$$

So you reason that you must switch envelopes if given the opportunity to do so, because the expected value, is trading your envelope for the other one, turned to be $\frac{5}{4}X$ is larger then X. But the same reasoning applies, have you been handed the other envelope, so something is definitely wrong because of the symmetry in the problem. This is the famous two-envelope paradox.

Due to the huge amount of articles and site (internet), there is no claim at all of having made a thorough review of literature on the subject. Looking at the web there are thousands of sites related to this problem. We looked at classical analysis given by Federico O'Reilly in 2010 which is very important to resolve the paradox. If you have X in your closed envelop, then the possibility that the other envelope has $\frac{X}{2}$ inside, since γ is fixed, means that your X=2γ, because the Benefactor has already decided the amount 3γ to be split in proportions 1:2 in the envelopes. On the other hand if you analyze the case where the other envelope has 2X, this means without any ambiguity that you have X= γ in your envelope. The expected value for the contents of the envelope you were given, depends on the fixed amount decided by the Benefactor, γ is given by

$$E(X;\gamma)=\frac{1}{2}\gamma+\frac{1}{2}2\gamma=\frac{3}{2}\gamma$$

If one analyzes the pseudo-expected value $E_X(\gamma) = \frac{5}{4}X$, this is meaningless. It has been incorrectly computed since it added the product of $\frac{1}{2}$ times $\frac{X}{2}$, the first summand with the product of $\frac{1}{2}$ times 2X, the second summand, so the correct expected value, substituting the quantity $\frac{X}{2}$ for its true value γ and the value 2X for its true 2γ is

$$E(Y;\gamma) = \frac{3}{2}\gamma$$

This last expected value coincides with the expected value of the contents of the envelope handed to you originally described in terms of γ. The incorrect use of the concept of expected value for the content in the second envelop, in terms of X, is responsible for the erroneous finding. The quantity γ is fixed and X is a quantity that varies according to the choice of the envelope, which in turn defines the two possibilities for Y. Now assume one of the following two decision principles of finite expectation maximization.

Maximin: For any two envelopes E_1 and E_2, if the minimal value of E_1 is greater than or equal to that of E_2, then you weakly prefer E_1 to E_2.

Maximax: For any two envelopes E_1 and E_2, if the maximal value of E_1 is greater than or equal to that of E_2, then you weakly prefer E_1 to E_2.

As mentioned above, indifference before opening can still be justified by the indifference principle. After opening, a maximaximizer will always switch, so that the two-envelope paradox remains for maximax preferences. A maximinimizer will keep after opening unless the observed amount is X=1. To obtain a two-envelope paradox also for maximinimizers, we drop the assumption that the amount x and y are powers of Z and assume instead that the pair (X,Y) has as its range the set of all pairs (X,Y) of rational number x and y such that

$$0 < x < 1 \text{ and } 0 < y < 1$$

where X and Y are two envelopes that have identical distribution i.e. that take each value with the same probability. As X and Y have the same (marginal) distribution. Then both maximinimizers and maximaximizers run into a two-envelope paradox, they are indifference before opening, but after opening a maximaximizer always switches and a maximinimizer always keep. This result is interesting in several respects (Chalmers, 2002).

- First, we have identified a two envelop paradox that involves no expectation reasoning whatsoever, neither before nor after opening.
- Second, an often criticized feature of the standard two-envelope paradox has been removed, namely the unboundedness of the possible payoff, both amounts range between 0 to 1 and of course have finite expectations.
- Third, we have removed the tight link between the two amount X and Y by allowing for a wide range of joint distributions of Xand Y, for instance for independence of X and Y.

O'Reilly (2006) discussed the Bayesian perspective to resolve the two-envelope paradox. In the original setting of the two-envelope paradox and more specifically in description of the expected value for the content of the other envelope in terms of the contents of the envelope handed to you, there appers to be agreement in most Bayesian papers dealing with this problem, in that is being implied is the conditional expectation of the contents of the other envelope given the contents of your envelope. The conditional expected value of Y given x may be given by

$$E(Y|x) = E\left(Y \middle| x, y = \frac{x}{2}\right) P\left(\gamma = \frac{x}{2} \middle| x\right) + E(Y|x, y = x) P(\gamma = x|x)$$

whereP(.$|x$) standing for the posterior distribution for γ. There is also agreement in that

$$E\left(Y \middle| x, y = \frac{x}{2}\right) = \frac{x}{2}$$

since under such conditions the distribution of Y is degenerate. Therefore, the expected value for the contents of the other envelope given x implies, without even making it explicit, that the posterior distribution of γ is uniform on the set $\{x|2, x\}$. Since indeed, E $(Y|x) = \frac{5}{4}x$ despite the symmetry in the setting of the problem. It is at this point that many authors mention that the problem lies in dealing with infinite expected values, for example when dealing with the improper prior, p (γ) $\alpha \frac{1}{\gamma}$.. It is fairly direct to recognize that because Y and X have exactly the same conditional expectation given γ, then follows taking expectations in both side, i.e.

$$E(X) = E(Y)$$

On the other hand,

$$E(Y|x) = \frac{3}{4}x$$

So again taking expectation both side, we get

$$E(Y) = \frac{3}{4}E(X)$$

These two quantities are thus true iff, both expected values are either 0or∞. This resolve the paradox.

Bibliography

1. Broome, J. (1995). The Two-Envelope Paradox, Analysis, 55(1), pp 6-11.
2. Bruss, F.T. (1996). The fallacy of the two-envelope problem, Mathematical Scientist, 21, pp 112-119.

3. Chalmers, D. (2002). The St. Petersburg Two-Envelope Paradox, Analysis, 62, pp155-157.
4. Christensen, R. and Utts, J. (1992), Bayesian Resolution of the Exchange Paradox, The American Statistician, 46, pp274-276.
5. Linzer, E. (1994). The Two-Envelope Paradox, The American Mathematical Monthly, 101, pp417-419.
6. O'Reilly, F. (2006). A look at the Two-envelope Paradox, STATS Issue, 45, pp 15-17.
7. Quintana, J.M., O'Reilly, F. (2008). Demystifying the Two-envelope Paradox?, Preimpreso No. 148. Septiem bre de 2008, IIMAS, UNAM, Mexico.
8. Ridgeway, T. (1993). Letter to the Editor about The Exchange Paradox, the American Statistician, 47, pp311.
9. Schwitzebel, www.faculty.ucr.edu/eschwitz/SchwitzAbs/TwoEnvelopSimple.htm

Chapter 42

Voting Paradox

A GROUP OF SEPARATELY RATIONAL INDIVIDUALS MAY HAVE PREFERENCES THAT ARE IRRATIONAL IN THE AGGREGATE.

The voting paradox, also known as Condorcet's paradox or the paradox of voting, is a situation noted by the "Marquis de Condorcet" in the late 18th century, in which collective preferences can be cyclic even if the preferences of individual voters are not. This is paradoxical, because it means that majority wishes can be in conflict with each other. When this occurs, it is because the conflicting majorities are each made up of different groups of individuals. Condorcet discovered that a group which uses majority rule to make decision can behave as if its preferences are intransitive even if every individual in the group has transitive preferences. If there are only two issues then majority rule works but if we have to choose between three or more issues. One method of making this choice is by pair wise comparisons each using majority rule. Suppose we have three candidates A, B and C and that there are three voters with preferences as follows

Voter	First Preference	Second Preference	Third Preference
Voter 1	A	B	C
Voter 2	B	C	A
Voter 3	C	A	B

If C is chosen as the winner, it can be argued that B should win instead, since two voters (1 and 2) prefer B to C and only one voter (3) prefer C to B. However, by the same argument A is preferred to B and C is preferred to A, by a margin of two to one on each occasion. The requirement of majority rule then provides no clear winner. Also, if an election were held with the above three voters as the only participants, nobody would win under majority rule, as it would result in a three way tie with each candidate getting one vote. A Condorcet method is a voting system that will always elect the Condorcet winner; this is the candidate whom voters prefer to each other candidate, when compared to them one at a time. This candidate can be found by conducting a series of pairwise comparisons, using the basic procedure described above. For N candidates, this requires $\frac{1}{2}$ N (N–1) pairwise hypothetical elections. For example, with 5 candidates there are 10 pairwise comparisons to be made. The several variants of the Condorcet method differ on how they resolve such ambiguities when they arise to determine a winner. Note that there is no fair and deterministic resolution to this trivial example because each candidate is in an exactly symmetrical situation.

As we talked earlier that the following set of preferences leads to a vote cycle. For understanding the implications of cycling in agenda setting consider the following example.

Voter	First Best	Second Best	Third Best
Voter 1	A	B	C
Voter 2	C	A	B
Voter 3	B	C	A

Suppose that instead of letting the vote cycle go on forever we use a set of rules which prescribes a certain order of voting and a limit on the number of votes. Under the random choice rule there are three possible agendas which are illustrated as

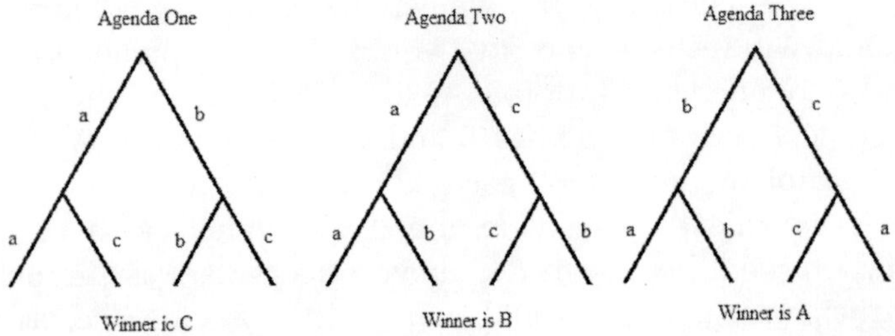

The arrows on the agendas indicates how voting proceeds given majority rule. In agenda two A is initially matched against C and C wins. In the beyond vote C is matched against B and is defeated. Notice that the agenda determines the winner. Preferences are the same in all three examples so the outcome is determined solely be something we would hope would be irrelevant the order of voting. In actual practice the random choice rule is rarely used. Instead the agenda is often controlled, at least to some extent, by one or more agents. In the Parliament House, for example, the majority party and the speaker of the House have significant control over the agenda. Of course, the minority party has some powers too and they also try to use their powers to control the agenda. Nevertheless, the majority party's power is significant. By choosing whether to use again, come to the problem, which one we vote on versus the other we get a different outcome, so we have a problem of vote cycling. If voting was efficient and consistent we should always get the same winner no matter which policy is put up against another. This problem with majority voting as it doesn't capture individual's performances. When we vote we have a problem in that the order in which we vote on alternatives does matter as to the outcome. This is the major flaw of majority voting if it was efficient we should always arrive at the same outcome, not different ones. This is why we see so many problems in the allocation of resources based on voting indeed the problem with allocation via voting is that it is impossible to devise a voting system that actually represents the underlying preferences of the voting public. This is called Arrow's Impossibility theorem in economics.

In the field of positive political theory, the term "impossibility theorem" refers to a collection of results that take as given a set of normatively appealing criteria and deductively prove that such a set of criteria are inconsistent, that is it impossible for a voting system, any voting system, to simultaneously satisfy the given set of criteria. This approach to studying voting systems is called axiomatic because the normative criteria that we might like our voting system to satisfy are used as starting points in these results, or taken as a given.

In other words, these normative criteria are axioms, which in statistics or mathematics are defined as statements that are assumed solely for the sake of studying the consequences that follow from them. The statistical or mathematical study of voting systems is motivated by the fact that any group seeking to make a collective decision must choose some method of translating the preferences of the group into a societal outcome. Moreover, there are an infinite number of such methods, from simple majority rule, to weighted voting, to choosing a dictator to unilaterally make societal decisions.

The formal statement of the Arrow's impossibility theorem can be understand by considering A, which is the set of outcomes, N a number of voters or decision criteria and L(A) denote the set of all full linear orderings to A. Then the social welfare function is given by

$$F : L(A)^N \to L(A)$$

which aggregates voter's preferences into a single preferences order on A. The N-tuple $(R_1, R_2, \ldots\ldots\ldots, R_N)$ of voter's preferences is called a preference profile. In its strongest and simplest form, Arrow's impossibility theorem states that whenever the set A of possible alternatives has more than two elements, the following three conditions become incompatible

Pareto Efficiency

If alternative 'a' is ranked above 'b' for all ordering $R_1, R_2, \ldots\ldots\ldots, R_N$ then 'a' is ranked higher than 'b' by

$$F(R_1, R_2, \ldots\ldots\ldots, R_N)$$

Non-Dictatorship:

There is no individual is whose preferences always prevail. That is, there is noi ? {1,2,......,N} such that

For all $(R_1, R_2, \ldots\ldots\ldots, R_N) \in L(A)^N$

andF $(R_1, R_2, \ldots\ldots\ldots, R_N) = R_i$

Independence of irrelevant alternative (IIA):

For two preference profiles $(R_1, R_2, \ldots\ldots\ldots, R_N)$ and $(S_1, S_2, \ldots\ldots.., S_N)$ such that for all individuals i, alternatives 'a' and 'b' have the same order in Ri as in Si alternative 'a' and 'b' have the same order in

$$F(R_1, R_2, \ldots\ldots\ldots, R_N) \text{ and } F(S_1, S_2, \ldots\ldots.., S_N)$$

There is another way to define the voting paradox using the impossibility theorem. This paradox demonstrates that majority rule with three or more alternatives may yield intransitive outcomes. It can be summed up by the following preference profile.

$$\rho = \begin{cases} x >_1 & y & >_1 z \\ Y >_2 & Z & >_2 x \\ Z >_3 & x & >_3 Y \end{cases}$$

where the preference relation is denoted $\geq_i$ for person i. Thus, if person i likes alternative x more than alternative y, it is denoted $x \geq i \ y$. Given profile ρ, it is clear that a majority vote between x and y would yield the social ranking x > y (by the votes of persons 1 and 3), a majority vote between y and z would yield y > z (by the votes of persons 1 and 2) and that a majority vote between x and z would yield the ranking z ? x (by the votes of persons 2 and 3). Thus, majority voting over pairs of alternatives is not guaranteed to yield transitive outcomes when there are three or more alternative, even when we assume that individual preferences are transitive. In this case, majority voting yields the cycle x > y, y > z and z > x.

Thus, in general, we can say that, this paradox is about the arguably irrational results that can arise in a collective choice among three or more alternatives. For the contention that an individual's vote will probably not effects the outcomes.

Bibliography

1. Arrow, K.J. (1950). A difficulty in the concept of social welfare, Journal of Political Economy, 58(4), pp 328-346.
2. Arrow, K.J. (1963). Social choice and individual value, Yale University Press, New Haven.
3. Austen-Smith, D. and Banks, J.S. (1999). Positive Political Theory I: Collective Preference, University of Michigan Press, Ann. Arbor.
4. Austen-Smith, D. and Banks, J.S. (2005). Positive Political Theory II: Collective Preference, University of Michigan Press, Ann. Arbor.
5. Balinski, M.L. and Young, H.P. (2001). Fair Representation: Meeting the ideal of one man one vote, Brooking Institute Press. Washington D.C.
6. Colomor, J (2004). The strategy and history of electoral system choice in Josep Colomer (ed.) Handbook of Electoral System choice, pp 3-73, Pourave Macmillan, New York.
7. Nurmi, H. (1999). Voting Paradox and How to Deal with Them, Springer-Verlag, Berlin.
8. Saari, D. (2001). Decisions and Elections. Explaining the Unexpected, Cambridge University Press, Cambridge.
9. Sen, A.K. (1970). Collective choice and social welfare, Holden-Day Sen Francisco.
10. Taylor, A.D. (2005) Social Choice and the Mathematics of Manipulation, Cambridge University Press, New York.